STATDISK
STUDENT LABORATORY MANUAL AND WORKBOOK

to accompany

Elementary Statistics
Eighth Edition

Mario F. Triola
Dutchess Community College

Addison
Wesley

Boston San Francisco New York
London Toronto Sydney Tokyo Singapore Madrid
Mexico City Munich Paris Cape Town Hong Kong Montreal

Trademarks

IBM and IBM PC are registered trademarks of International Business Machines Corporation.

Macintosh is a registered trademark of Apple Computer, Inc.

MINITAB is a registered trademark of MINITAB, Inc.

Addison Wesley Longman Publishing Company, Inc. and the author make no representations or warranties, express or implied, with respect to this software and its documentation, including without limitations, any implied warranties of merchantiblity, or fitness for any particular purpose, all of which are expressly disclaimed. The exclusion of implied warranties is not permitted in some states. The above exclusion may not apply to you. This warranty provides you with specific legal rights. There may be other rights that you have which may vary from state to state.

Reproduced by Addison-Wesley Publishing Company Inc. from camera-ready copy supplied by the authors.

ISBN   0-201-70466-8

1 2 3 4 5 6 7 8 9 10 PHTH 03 02 01 00

Preface

This *STATDISK Student Laboratory Manual and Workbook,* 8th edition, and the new STATDISK Version 8.1, are supplements to the 8th edition of *Elementary Statistics* by Mario F. Triola. The STATDISK software, included with the CD-ROM that is packaged with the textbook, is designed for IBM PC and compatible computers using Windows, as well as Macintosh computers. For updates to the software, visit the Web site at

http://www.awl.com/triola

This manual/workbook generally refers to the Windows version of the software, and there are some differences between this version and the Macintosh version. For example, you can exit the Windows version by clicking on File, then Exit; you can exit the Macintosh version by clicking on File, then Quit.

STATDISK Version 8.1 has new features not included in previous versions. Section 5-7 of *Elementary Statistics* is a new section that introduces normal quantile plots as a tool to help in determining whether a sample comes from a normally distributed population. STATDISK Version 8.1 has normal quantile plots as a new feature. The Normal Distribution module allows you to get probabilities associated with the normal distribution. The previous version of STATDISK was programmed so that the Normal Distribution module allowed only a mean of 0 and a standard deviation of 1, but the new version allows you to enter any mean and any positive standard deviation. In past versions of STATDISK, printing large data sets was not easy. The new version has been modified to simplify that process. (See Section 1-7 of this manual/workbook.) When conducting hypothesis tests involving two means from small and independent samples, previous versions of STATDISK automatically applied an *F* test to determine whether the sample variances should be pooled. STATDISK Version 8.1 now gives you the option of pooling the sample variances, not pooling the sample variances, or using a preliminary *F* test. Where the textbook changed terminology, the corresponding terms in STATDISK were changed accordingly. For example, instead of referring to two dependent samples, the textbook now refers to "matched pairs" of data, and STATDISK has also been changed to use the "matched pairs" terminology. The data sets included with STATDISK have been updated for the 8th edition of the textbook.

STATDISK is intended to be an educational tool that supplements *Elementary Statistics*, not a commercial tool like SPSS, or SAS, or Minitab. STATDISK and this manual/workbook were developed specifically for those instructors who wish to incorporate computer usage, but recognize that limitations of time do not allow extensive computer instruction. Instructors can assign computer experiments from this manual/workbook without using the valuable class time that is already quite limited. Although this manual/workbook is not intended to describe a comprehensive course in statistics, it allows the student to further explore topics and discover important principles by conducting computer experiments. It is assumed that the necessary foundation for such exploration and discovery are the principles and concepts presented in *Elementary Statistics*, 8th edition.

Chapter 1 of this supplement describes some of the important basics for using STATDISK. Chapters 2 through 13 in this manual/workbook correspond to Chapters 2 through 13 in *Elementary Statistics*, 8th edition. However, the individual chapter *sections* in this manual/workbook generally do *not* match the sections in the textbook. (The sections for Chapters 8 and 13 in this manual/workbook do match the sections in the textbook.) Each chapter includes a description of the STATDISK procedures relevant to the corresponding chapter in the textbook. This cross-referencing makes it very easy to use this supplement with the textbook. Except for Data Set 6, STATDISK also includes the data sets found in Appendix B of the textbook. Those data sets are already available in the STATDISK program and there is no need to import them.

Chapters include a beginning section in which examples are illustrated with sample runs of STATDISK. It would be helpful to follow the steps shown in these sections so the basic procedures will become familiar. You can compare your own computer display to the display given in this supplement and then verify that your procedure works correctly. You can then proceed to conduct the experiments that follow.

I wish to thank Bill Flynn for his expertise and competency in developing the original STATDISK algorithms. For this new version of STATDISK, I thank Stacie Wogalter of Tallent Technology Corporation. For the current installation program, I thank Tony Denizard, a friend and former student. I also thank Russell F. Loane and Timothy C. Armstrong of Password, Inc. for their truly exceptional work on the previous version. Their dedication and talent are very apparent in the current version of STATDISK. Finally, I thank the Addison Wesley Longman staff for their enthusiastic support in this project. It is a genuine pleasure working with a publishing company committed to providing a product with the highest quality. I also thank the many instructors and students who took the time to provide many valuable suggestions.

We welcome any comments or suggestions for improving STATDISK or this manual/workbook. Please send them to the Addison Wesley Longman Statistics Editor.

M.F.T.
May, 2000

Contents

1

STATDISK
Fundamentals

STATDISK is designed so that it uses many of the same features common to a wide variety of software applications, so tools introduced in this chapter have a universal usefulness that extends beyond STATDISK and statistics. For example, the **Copy** and **Paste** features of STATDISK are commonly included with many software programs. As you learn how to use such features, you acquire important and general computer skills that will help you with other applications.

1-1 Installing STATDISK

The STATDISK statistics software is on the CD-ROM that is included with *Elementary Statistics*. In some cases, your instructor will have STATDISK installed on your college network. If the network is used, your instructor will provide access instructions. If a network is not used, you can install your own copy on your own computer.

> **IBM PC compatible computers with Windows**: Put the CD-ROM in its disk drive, click on Start, then Run, and enter the drive containing the CD-ROM followed by :/Software/Statdisk/Setup.exe. For example, if your CD-ROM is in drive D, click on **Start**, then **Run**, then enter **D:/Software/Statdisk/Setup.exe**.

> **Macintosh computers**: Copy the STATDISK program from the CD-ROM to the computer's hard drive by clicking and dragging the program icon. Start the program by double-clicking on the STATDISK icon. The CD-ROM location is Software/Statdisk. (Note: If your CD-ROM has STATDISK version 7, download the latest version 8 from the Web site http://www.awl.com/triola.)

1-2 Entering Data

Your first use of STATDISK is likely to occur with topics from Chapter 2 of *Elementary Statistics*, and one of your first objectives is likely to be entering a set of sample data. (Almost all of the sample data found in Appendix B of *Elementary Statistics* are already stored with STATDISK and they can be retrieved as described in Section 1-4 of this manual/workbook. It is not necessary to manually enter those data sets that are already stored with STATDISK.) To manually enter a set of sample data, use the following procedure.

STATDISK Procedure for Entering Data

1. The main menu bar contains the items shown below.

   <u>F</u>ile   <u>E</u>dit   <u>A</u>nalysis   **Data**   <u>H</u>elp

   Use the mouse to click on **Data**. (Instead of using the mouse, you can also press Alt-D, but the mouse is usually easier.)

2. In the menu that pops up, click on the first item of **Sample Editor**.

3. You should now see the box shown on the top of the next page.

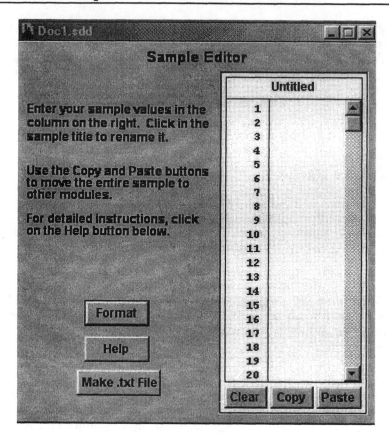

In the above box, the cursor will be blinking in the position for the first data value. Type your first data value, then press the **Enter** key. Then type the second data value and press the **Enter** key again. Proceed to type a sample value followed by pressing the **Enter** key until all of the sample values have been entered. Although the above STATDISK display shows only 20 positions, you can enter up to 1000 values. *Note*: If you see that you have made a mistake, simply click on the wrong value and press the **Del** key or **Backspace** key to erase it, then type the correct value in its place.

4.   *Name the data set* if you plan to save it. Click on the title of "Untitled" at the top of the data set listing, use the **Del** key or **Backspace** key to delete "Untitled" and type the name you have chosen for the data set.

There are a few other important features that are available by clicking on the **Format** bar. When in the Sample Editor window, if you click on the Format bar you will get a screen like the one shown on the following page. Here are useful features of that screen:

Num columns: You can increase the number of columns of data from 1 up to 9.

Num decimals: You can enter the number of decimal places appropriate for your data set. For example, if you are entering grade point averages such as 3.125, 2.972, and 2.058, you should set the number of decimal places to 3. (If you enter data without specifying the number of decimal places, STATDISK will set that number based on the values that are entered.)

Sort Order: Select "none" if you want the values listed as they are entered, or select "ascending" if you want the data sorted (arranged in order) from lowest to highest, or select "descending" if you want the data arranged in order from highest to lowest.

The **Make .txt File** bar at the bottom of the Sample Editor screen is used to convert the data set from a STATDISK file to a text file with extension .txt. See Sections 1-7 and 1-9 for more information about this bar.

After entering a large data set, you often want to save it for future use, and you often want to use the data in other modules, such as the modules designed to generate histograms, boxplots, or descriptive statistics (all described in Chapter 2 of the textbook). Be sure to see Section 1-3 (for saving data) and Section 1-5 (for using *copy* and *paste*).

1-3 Saving Data

After entering a set of data as described above, you can save it for future use by the following procedure.

STATDISK Procedure for Saving Data

1. After entering all of the values in the Sample Editor dialog box, use the mouse to click on **File** located on the main menu.

    File   Edit   Analysis   Data   Help

2. A menu will pop up. Click on the third item of **Save As....**

3. You will now see a dialog box with the title of "Save File As." A title such as Doc1.sdd will be in the entry box, so click on that box, press Del (delete) to delete that default label, and enter a name for the data set; the extension of .sdd will be automatically affixed to the end of the file name, so that it will be stored

as a genuine STATDISK file, instead of some nondescript file that doesn't know where it belongs. (Shown below is the "Save As..." dialog box with the file name of bodytemp and settings for saving the file in the STATDISK file of drive C.) Click **OK** and the data set will be saved under the name you assigned.

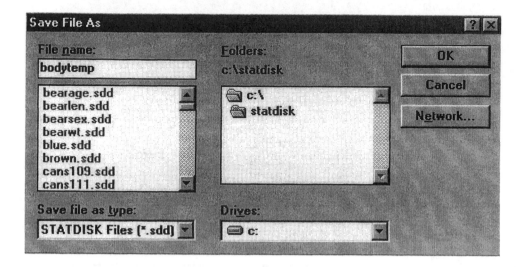

If you want to save your file in a location different from the location in which STATDISK resides, you can change the drive and folder as you desire. For example, if you have a floppy disk in drive A and you want to save the data set there, you can change to drive to A by clicking on the box in the lower right corner of the "Save As..." dialog box.

1-4 Retrieving Data

STATDISK comes with 97 data sets already stored. See Appendix B in *Elementary Statistics* for the file names. Some Exercises in the textbook require that you use some of these data sets. Also, you may want to retrieve a data set that you have previously entered and saved. To retrieve one of the stored data sets, follow these steps.

1. Click on the main menu item of **File**.

2. Click on the subdirectory item of **Open**.

3. You should now see a window like the one shown on the next page.

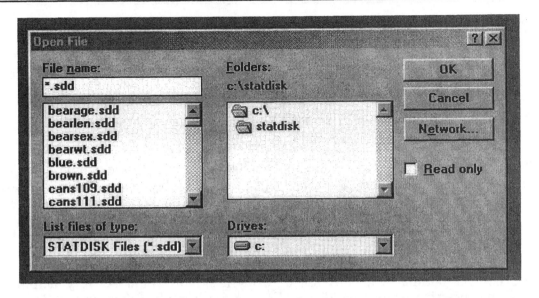

There are many more data sets stored than the 8 shown in this window. You can scroll through this list using your mouse and the up-arrow and down-arrow buttons located to the right of the list. (To scroll faster, click on the small bar just below the up-arrow button and, while holding the mouse button down, slide the mouse downward.)

4.   After finding the data set you want to retrieve, click on its name.

5.   Click the **OK** button, and the data set will be listed in the Sample Editor window, so you can now use the copy and paste buttons (see Section 1-5) to move the data into the module you would like to use.

1-5 Copy/Paste

The Copy and Paste feature is an extremely important tool of many different software applications, including word processors and spreadsheets. By using the copy and paste buttons, you can take a data set and move a copy of it to another module where you can do something with it. You should clearly understand the following.

- **After entering or retrieving a data set and clicking on the *Copy* button, the data set will remain available for use until you click on *Copy* for a new data set, or until you exit the program.**

- **After clicking on *Copy*, go to the module where you want to use the data set, then click on the *Paste* button in the window for that application.**

If you have a data set in the Sample Editor dialog box, you can transfer the data to other modules and do wonderful things. For example, suppose you want to enter a data set and obtain descriptive statistics, including statistics such as the mean, median, standard deviation, quartiles, and sum of the values (all of which are discussed in Chapter 2 of the textbook). Use the following procedure.

STATDISK Procedure for Using Copy and Paste

1. Enter the data in the Sample Editor module (as described in Section 1-2 above) or retrieve a stored data set into the Sample Editor module (as described in Section 1-4 above).

2. Use the mouse to click on the **Copy** button located at the bottom of the Sample Editor window.

3. You can now go to any module that uses a data set. For example, click on **Data** on the main menu bar, then click on the subdirectory item of **Descriptive Statistics**. Click on the **Paste** button located at the bottom of the Descriptive Statistics window and you will see the original data set reappear in this Descriptive Statistics window. You can now click on the Evaluate button to obtain the descriptive statistics for the data set.

If you are not yet familiar with this use of Copy/Paste, it would be very helpful to load STATDISK and practice using Copy/Paste. Enter a few sample values in the Sample Editor window, then use Copy/Paste to get the descriptive statistics for the data by using the above procedure.

1-6 Editing and Transforming Data

It is easy to edit a data set while in the Sample Editor module.

- *Delete* an entry by using the **Del** key. (You can delete whole sections of data by holding the mouse button and dragging the mouse to highlight the block of values, then select Edit from the main menu, and click on "Delete All.")

- *Insert* an entry by typing it in the desired location, then press the Enter key.

Data may also be *transformed* with operations such as adding a constant, multiplying by a constant, or using functions such as logarithm (common or natural), sine, exponential, or absolute value. For example, if you have a data set consisting of temperatures on the Fahrenheit scale (such as Data Set 6 in Appendix B of the textbook) and you want to transform the values to the Celsius scale, you can use the equation

$$C = \frac{5}{9}(F - 32)$$

STATDISK Procedure for Transforming Data

1. Click on the main menu item of **Data**.

2. Click on **Sample Transformations** to get a window like the one shown on the next page.

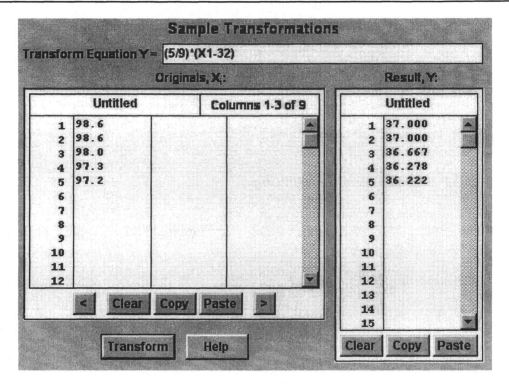

3. Either manually enter the sample data in the leftmost column, or use Copy/Paste to move a data set into the Sample Transformations window. (If you use Copy/Paste, click on the Paste bar in the left portion of the window.)

4. Enter an expression in the top window. The above screen display shows that the sample data are entered as 98.6, 98.6, 98.0, 97.3, and 97.2. These values are body temperatures in degrees Fahrenheit. They are converted to the Celsius scale by using the equation entered as Y= (5/9)*(X1 - 32).

5. Click on **Transform** and the transformed values will appear in the column at the extreme right. The five values in the extreme right column of the above display are the Celsius equivalent values of the listed Fahrenheit body temperatures. You can now save or copy the transformed data as you desire.

For more details of the procedure for transforming data, select Data from the main menu, then select Sample Transformations, then click on the Help bar.

1-7 Printing Screens and Data Sets

After you have successfully obtained results from a module, such as a graph or a listing of statistics, you can print the window.

Printing Screens

1. Use the mouse to click on the main menu item of **File**.

2. Click on the subdirectory item of **Print**.

After entering and saving a data set, there is often a need to also obtain a printed copy of it. Because the Sample Editor window displays only 20 sample values at a time, it is not convenient for printing large data sets. There are other procedures for printing large data sets.

Printing a Large Data Set in a Word Processor

1.   With the data set displayed in the Sample Editor window, click on the value at the top. Hold the mouse button down and drag the mouse to the bottom of the data set, then release it. The entire list of values should be highlighted.

2.   Click on **Edit**, then click on **Copy**.

3.   Now go into your word processor and click on **Edit**, then **Paste**. The entire list of values will be in your word processing document where you can configure them as you please.

Printing a Large Data Set in STATDISK

1.   With the data set displayed in the Sample Editor window, click on the **Make .txt File** bar at the bottom. (This will create a text file with extension .txt.)

2.   In the window that appears, enter a file name or accept the default name.

3.   Click on the main menu item of **Data**, then click on the last item of **Print Screen via File/Print**.

4.   Click on the **Import Text** bar in the upper left corner.

5.   In the Open File window that appears, click on the name of the file that you want displayed, then click **OK**. The data set will appear in the current window.

6.   You can print separate pages until the entire data set has been printed, but you can also use the **End**, **Del**, and arrow keys to reconfigure the data to a more convenient format. For example, this approach was used to print on *one* page the 175 values in the STATDISK file cans109.sdd (see Data Set 12 in Appendix B) with the result shown below. You can change the number of columns as you desire.

270 273 258 204 254 228 282 278 201 264 265 223 274 230 250 275 281 271 263 277
275 278 260 262 273 274 286 236 290 286 278 283 262 277 295 274 272 265 275 263
251 289 242 284 241 276 200 278 283 269 282 267 282 272 277 261 257 278 295 270
268 286 262 272 268 283 256 206 277 252 265 263 281 268 280 289 283 263 273 209
259 287 269 277 234 282 276 272 257 267 204 270 285 273 269 284 276 286 273 289
263 270 279 206 270 270 268 218 251 252 284 278 277 208 271 208 280 269 270 294
292 289 290 215 284 283 279 275 223 220 281 268 272 268 279 217 259 291 291 281
230 276 225 282 276 289 288 268 242 283 277 285 293 248 278 285 292 282 287 277
266 268 273 270 256 297 280 256 262 268 262 293 290 274 292

1-8 Closing Windows and Exiting

Closing Windows: As you use STATDISK, you will be opening various windows. It is usually wise to *close* windows after they are no longer needed. (Because of memory limitations, there is a limit to the number of data sets that may be opened.) To keep memory available and to keep clutter to a minimum, close windows by clicking on the small box labeled × that is located in the upper right corner of the window. (Clicking on the box with the symbol – will cause the window to be hidden, but it continues to remain open and available for recall.)

Exiting STATDISK: Had enough for now? To exit or quit the STATDISK program, click on the × located in the extreme upper right corner. Another way to exit STATDISK is to click on **File**, then click on **Exit**.

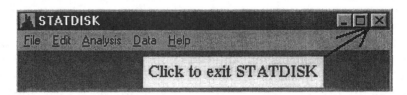

1-9 Exchanging Data with Other Applications

There may be times when you want to move data from STATDISK to another application (such as Excel or Minitab or Word) or to move data from another application to STATDISK. Instead of manually retyping all of the data values, you can usually transfer the data set directly. Given below are two ways to accomplish this. The first of the following two procedures is easier, if it works (and it often does work).

Method 1: Use Edit/Copy and Edit/Paste

1.   With the data set displayed as a column in the data source, click on the value at the top. Hold the mouse button down and drag the mouse to the bottom of the data set, then release it. The entire list of values should be highlighted.

2.   Click on **Edit**, then click on **Copy**.

3.   Now go into the software program that is your destination and click on **Edit**, then **Paste**. The entire list of values should reappear.

Method 2: Use Text Files

1.   In the software program containing the original set of data, create a text file of the data. (In STATDISK you can use the **Make .txt File** bar on the bottom of the Sample Editor window, or you can use **Data** and **Sample Converter**.)

2.   STATDISK and most other major applications allow you to import the text file that was created. (To import a text file into STATDISK, select **Data**, then **Sample Converter**. Click on **Import Text**, select the desired text file, then click **OK**, then click **Text to Sample** and the data set is available as a STATDISK file.)

CHAPTER 1 EXPERIMENTS: STATDISK Fundamentals

1-1.   **Entering Sample Data**   When first experimenting with procedures for using STATDISK, it's a good strategy to use a small data set instead of one that is large. If a small data set is lost, you can easily enter it a second time. In this experiment, we will enter a small data set, save it, retrieve it, and print it. Table 6-1 in *Elementary Statistics* includes these body temperatures, along with others:

98.6   98.6   98.0   98.0   99.0   98.4   98.4   98.4   98.4   98.6

a.   Load STATDISK and enter the above sample values. (See the procedure described in Section 1-2 of this manual/workbook.)

b.   Save the data set using the file name of TEMP (because the values are body temperatures). See the procedure described in Section 1-3 of this manual/workbook.

c.   Print the data set by printing the display of the Sample Editor window.

d.   Exit STATDISK, then reload it and retrieve the file named TEMP. Save another copy of the same data set using the file name of TEMP2. Print TEMP2 and include the title at the top. (That is, use a title of TEMP2 instead of the default name of UNTITLED.)

1-2.   **Entering a Large Data Set**   Experiment 1-1 involves only 10 sample values. Enter the complete data set of 106 values reproduced below, then save the data set with the name of BODYTEMP. Return to the Sample Editor window, click on **Format,** and proceed to sort the values in ascending order. Print a copy of the sorted data on *one* page (see Section 1-7).

98.6	98.6	98.0	98.0	99.0	98.4	98.4	98.4	98.4	98.6
98.6	98.8	98.6	97.0	97.0	98.8	97.6	97.7	98.8	98.0
98.0	98.3	98.5	97.3	98.7	97.4	98.9	98.6	99.5	97.5
97.3	97.6	98.2	99.6	98.7	99.4	98.2	98.0	98.6	98.6
97.2	98.4	98.6	98.2	98.0	97.8	98.0	98.4	98.6	98.6
97.8	99.0	96.5	97.6	98.0	96.9	97.6	97.1	97.9	98.4
97.3	98.0	97.5	97.6	98.2	98.5	98.8	98.7	97.8	98.0
97.1	97.4	99.4	98.4	98.6	98.4	98.5	98.6	98.3	98.7
98.8	99.1	98.6	97.9	98.8	98.0	98.7	98.5	98.9	98.4
98.6	97.1	97.9	98.8	98.7	97.6	98.2	99.2	97.8	98.0
98.4	97.8	98.4	97.4	98.0	97.0				

1-3.   **Retrieving Data**   The file CANS109 is already stored in STATDISK. It contains 175 values that are listed in Data Set 12 in Appendix B of *Elementary Statistics*. Open that file and print the data on *one* page (see Section 1-7).

1-4.   **Using Copy/Paste**   Enter the sample values 1, 2, 5, 6, 8, 8 and use Copy/Paste to copy the data set to the Descriptive Statistics module (which is accessible from the **Data** menu). After pasting the data set to the Descriptive Statistics module, click on **Evaluate** and obtain a printed copy of the resulting screen display. The resulting statistics will be described in Chapter 2 of the textbook.

1-5.   **Using Copy/Paste**   The file NICOTINE is already stored in STATDISK. Open that file, then use Copy/Paste to copy it to the Descriptive Statistics module (which is accessible from the **Data** menu). After pasting the data set to the Descriptive Statistics module, click on **Evaluate** and obtain a printed copy of the resulting screen display. The resulting statistics will be described in Chapter 2 of the textbook.

1-6. **Editing Data** The file MCGWIRE lists the 70 distances of the homeruns hit by Mark McGwire when he broke a major baseball record.
    a.   Open the file MCGWIRE, then use Copy/Paste to copy it to the Descriptive Statistics module. Click on Evaluate and record the value of the mean._____
    b.   Go back to the Sample Editor and change the fifth value from 420 to 4200. Repeat part a and record the new value of the mean._____
    c.   Did the mean change much when the fifth value was changed from 420 to 4200?

_____

1-7. **Generating Random Data** In addition to entering or retrieving data, STATDISK can also *generate* data sets. In this experiment, we will use STATDISK to simulate the rolling of a pair of dice 500 times.
    a. Select **Data** from the main menu bar.
    b. Select **Dice Generator**.
    c. For the sample size, enter 500 (for 500 rolls).
    d. Enter 2 for the number of dice.
    e. Enter 6 for the number of sides.
    f. Click on Generate.
    g. Examine the displayed totals and count the number of times that 7 occurs. Record the result here: _____

1-8. **Transforming Data** Experiment 1-2 results in saving the data set BODYTEMP that contains 106 body temperatures in degrees Fahrenheit. Retrieve that data set, then proceed to transform the temperatures to the Celsius scale. (See Section 1-6 in this manual/workbook.) After getting the Celsius temperatures in the rightmost column of the Sample Converter window, use Copy/Paste to copy the transformed data to the Descriptive Statistics module (accessible through the main menu item of **Data**). Click on **Evaluate** and record the value of the mean here:_____

1-9. **Retrieving Data** Data Set 1 in Appendix B of the textbook includes the weights (in pounds) of a sample of 36 cans of regular Coke. Those weights are stored in the STATDISK file CKREGWT. Copy those weights to the Descriptive Statistics module and record the following statistics.

Mean:_____   Minimum:_____   Maximum:_____

1-10. **Retrieving and Transforming Data** Retrieve the STATDISK file BEARWT, which lists the weights of a sample of bears. Those weights are in pounds. To convert the weights to kilograms, multiply them by 0.4536. Use STATDISK to convert the weights from pounds to kilograms. In the space below, write the weights (in kilograms) of the first five bears.

_____

2

Describing, Exploring, and Comparing Data

Important note: The topics of this chapter require that you use STATDISK to enter data, retrieve data, save files, and print results. These functions are covered in Chapter 1 of this manual/ workbook. Be sure to understand these functions before beginning this chapter.

The main objective of Chapter 2 in *Elementary Statistics* is to introduce the tools needed to describe, explore, or compare those characteristics of a data set that are extremely important.

Important Characteristics of Data

When describing, exploring, and comparing data sets, the following characteristics are usually extremely important:

1. **Center:** Measure of center, which is a representative or average value that gives us an indication of where the middle of the data set is located

2. **Variation:** A measure of the amount that the values vary among themselves

3. **Distribution:** The nature or shape of the distribution of the data, such as bell-shaped, uniform, or skewed

4. **Outliers:** Sample values that are very far away from the vast majority of the other sample values

5. **Time:** Changing characteristics of the data over time

In this chapter, we learn how to use STATDISK as a tool for investigating the above important characteristics.

The Chapter Problem in Chapter 2 addresses this question: "Is there a keyboard configuration that is more efficient than the one that most of us now use?" The traditional keyboard configuration is called a *Qwerty* keyboard because of the positioning of the letters QWERTY on the top row of letters. The Dvorak keyboard supposedly provides a more efficient arrangement by positioning the most used keys on the middle row (or "home" row) where they are more accessible. Chapter 2 in the textbook used a rating system with each of the 52 words in the Preamble to the Constitution to obtain Tables 2-1 and 2-2.

Table 2–1	**Qwerty Keyboard**								
	Word Ratings								
2	2	5	1	2	6	3	3	4	2
4	0	5	7	7	5	6	6	8	10
7	2	2	10	5	8	2	5	4	2
6	2	6	1	7	2	7	2	3	8
1	5	2	5	2	14	2	2	6	3
1	7								

Table 2–2	**Dvorak Keyboard**								
	Word Ratings								
2	0	3	1	0	0	0	0	2	0
4	0	3	4	0	3	3	1	3	5
4	2	0	5	1	4	0	3	5	0
2	0	4	1	5	0	4	0	1	3
0	1	0	3	0	1	2	0	0	0
1	4								

Interpreting Results: We noted in the textbook that a visual comparison of Tables 2-1 and 2-2 might not reveal very much. We now proceed to use STATDISK to gain meaningful insight into such data sets.

2-1 Descriptive Statistics

To obtain descriptive statistics (mean, median, standard deviation, and so on) for a set of sample values, follow these steps.

1. Enter or retrieve a set of sample data. (To *enter* values, use **Data/Sample Editor** as described in Section 1-2 in this manual/workbook; to *retrieve* a data set, use **File/Open** as described in Section 1-4 of this manual/workbook.) After entering or retrieving a data set, you will be in the **Sample Editor** window.

2. Click on the **Copy** bar at the bottom of the Sample Editor window. (See Section 1-5 of this manual/workbook.)

3. Click on **Data** in the main menu bar at the top.

4. Click on **Descriptive Statistics**.

5. Click on the **Paste** bar at the bottom. The data should appear in the window.

6. Click on the **Evaluate** bar located on the bottom left side of the window.

As an example, *enter* the 52 QWERTY values listed in Table 2-1. Enter these values by first clicking on **Data**, then selecting **Sample Editor**. Proceed to enter the first value of 2, then press **Enter**. Now enter the second value of 2 and press **Enter**. Enter the third value of 5 and press **Enter**, and continue until all 52 values have been entered. Take great care to enter the values correctly. With the data listed in the Sample Editor window, continue with Step 2 in the above procedure. The data should now be copied in the Descriptive Statistics module. Click on Evaluate, and the window display should be as shown below.

Sample Descriptive Statistics			
Untitled		**Untitled**	
Sample Size, n	52	1	2
		2	2
Mean, x̄	4.4038	3	5
Median	4.0000	4	1
Midrange	7.0000	5	2
RMS	5.2275	6	6
		7	3
Variance, s^2	8.0886	8	3
St Dev, s	2.8440	9	4
Mean Dev	2.3425	10	2
Range	14.000	11	4
		12	0
Minimum	0.0000	13	5
1st Quartile	2.0000	14	7
2nd Quartile	4.0000	15	7
3rd Quartile	6.0000	16	5
Maximum	14.000	17	6
		18	6
Σx	229.00	19	8
Σx^2	1421.00	20	10

Evaluate   Help   Clear   Copy   Paste

From the above STATDISK display, you can see the values of important descriptive statistics. The value of "RMS" is the value of the root mean square (or quadratic mean) described in Exercise 19 of Section 2-4 in *Elementary Statistics*. The value listed as "Mean Dev" is the mean deviation (or mean absolute deviation) defined in Section 2-5. Section 2-7 of the textbook includes the definition of a "5-number summary (minimum, 1st quartile, 2nd quartile, 3rd quartile, maximum), and that summary is located near the bottom of the STATDISK display.

Based on the items included in the above STATDISK display, we now have an understanding of some of these important characteristics of data:

Center:	The mean is 4.4 and the median is 4.0 (rounded).
Variation:	The standard deviation is 2.8, the variance is 8.1, and the range is 14.0.

To understand other important characteristics of data, such as distribution, outliers, and pattern over time, we need other tools of STATDISK. In the following section of this manual/workbook, we consider histograms and frequency tables, which give us important insight into the nature of the distribution of the data.

2-2 Histograms and Frequency Tables

In designing STATDISK, we did not include a specific menu item for generating a frequency table, but frequency tables can be obtained by using the ability of STATDISK to generate a histogram. Section 2-3 of the textbook describes the manual construction of a histogram, but STATDISK can be used to automatically generate this important graph. The basic approach is to enter or retrieve a data set, then use Copy/Paste to move the data to the Histogram module, where the histogram graph is generated. When using STATDISK's Histogram program, you have the option of simply accepting default settings, or you can set your own limits on the different classes. If you choose to set your own limits, you must understand the definition of *class width*. In *Elementary Statistics*, we defined class width as follows:

> **Class width** is the difference between two consecutive lower class limits or two consecutive lower class boundaries.

Let's return to this definition after describing the STATDISK procedure for generating a histogram.

Procedure for Generating a Histogram

1.   Enter or retrieve a set of sample data. (To *enter* values, use **Data/Sample Editor** as described in Section 1-2 in this manual/workbook; to *retrieve* a data set, use **File/Open** as described in Section 1-4 of this manual/workbook.) After entering or retrieving a data set, you will be in the **Sample Editor** window.

2.   Click on the **Copy** bar at the bottom of the Sample Editor window. (See Section 1-5 of this manual/workbook.)

3.   Click on **Data** in the main menu bar at the top.

4.   Click on **Histogram**.

5.   Click on **Paste**. (The sample data should now be listed in the Histogram window.)

6.   Click on **Evaluate** (or click on Help if you first want additional information about generating a histogram).

7.   A small window will appear with the title of Histogram Options. The easy way to proceed is to simply click **OK**. (You have the option of changing the class width and starting point of the first class. You also have the option of having a vertical scale that uses relative frequencies instead of actual frequency counts. Click OK after making the desired changes.)

If we use the 52 Qwerty values in Table 2-1 and accept the STATDISK default settings, the histogram will be as shown on the next page.

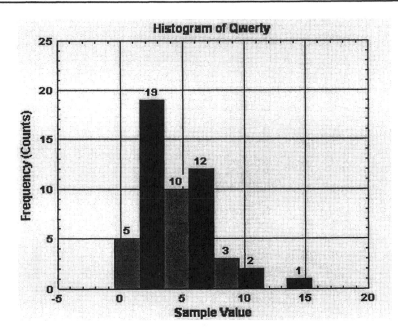

The above histogram was generated using the default settings shown in this STATDISK dialog box:

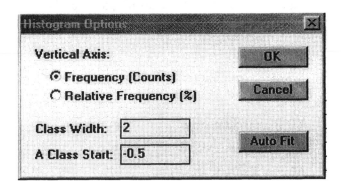

You can see that for this particular data set, the default class width is 2 and the first class begins at -0.5. Using those values, the histogram is based on these class boundaries:

$$-0.5 - 1.5$$
$$1.5 - 3.5$$
$$3.5 - 5.5$$
$$5.5 - 7.5$$
$$7.5 - 9.5$$
$$9.5 - 11.5$$
$$11.5 - 13.5$$
$$13.5 - 15.5$$

The frequencies for the different classes are located above the bars of the histogram.

If, instead of accepting the STATDISK default settings, we change the class width to 3 and we change the class start to 0, we get the following histogram.

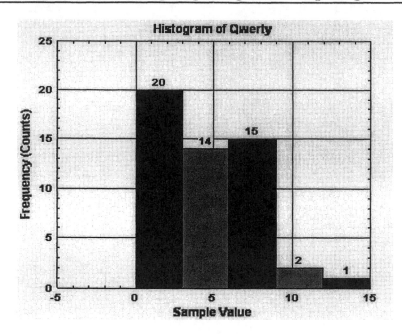

Given that the above histogram was generated with a class width of 3 and a starting point of 0, and given that all of the sample values are whole numbers, the above histogram can be used to construct this frequency table, which is Table 2-4 in the textbook:

x	f
0 – 2	20
3 – 5	14
6 – 8	15
9 – 11	2
12 – 14	1

When using STATDISK's Histogram program, it is easy to accept the program defaults, which is fine if your sole objective is to see a graph of the *distribution* of the data. If you want to use STATDISK to construct a frequency table, choose the option of entering your own starting point and class width (based on the range of values and the minimum value).

Among the important characteristics of data, the histogram gives us insight into the nature of the *distribution*. In later chapters, it often becomes important to determine whether sample data appear to come from a population with a normal distribution, and that determination can often be made by visual examination of a histogram. We simply examine the graph and make a judgment about whether it appears to be approximately bell-shaped. If we examine both of the STATDISK histograms shown in this section, we can see that the distribution appears to be somewhat heavier to one side (or *skewed*), so that the requirement of a normal distribution does not appear to be satisfied.

2-3 Boxplots

Section 2-7 of *Elementary Statistic* describes the construction of boxplots. They are based on the 5-number summary consisting of the minimum, first quartile, second quartile, third quartile, and maximum. The basic approach is to enter or retrieve a data set, then use Copy/Paste to copy the data to the Boxplot module, which is one of the modules under the main menu item of Data.

Procedure for Generating a Boxplot

1. Enter or retrieve a set of sample data. (To *enter* values, use **Data/Sample Editor** as described in Section 1-2 in this manual/workbook; to *retrieve* a data set, use **File/Open** as described in Section 1-4 of this manual/workbook.) After entering or retrieving a data set, you will be in the **Sample Editor** window.

2. Click on the **Copy** bar at the bottom of the Sample Editor window. (See Section 1-5 of this manual/workbook.)

3. Click on **Data** in the main menu bar at the top.

4. Click on **Boxplot**.

5. Click on **Paste**.

6. In the window for selecting a destination, enter column 1, then click **OK**.

7. Click on **Evaluate**.

8. In the window for boxplot options, you must select the columns to plot. (You have the option of plotting several boxplots in the same window.) If there is only one data set, select column 1, then click **OK**.

Chapter 2 in the textbook notes that one important advantage of boxplots is that they are very useful in comparing data sets. Shown below is the STATDISK display showing the two boxplots representing the two sets of sample data summarized in Table 2-1 (Qwerty Word Ratings) and Table 2-2 (Dvorak Word Ratings). To get more than one boxplot displayed in the same window, follow steps 1-5 above to enter the data for the first set of values. Then enter or retrieve another data set and use Copy/Paste to copy it to the *same* boxplot window, but enter the second set of data in column 2. Continue until you are finished entering all of the data sets (up to 9), then continue with steps 7 and 8 above.

In the display shown below, the top boxplot depicts the Qwerty word ratings and the bottom boxplot represents the Dvorak word ratings. Because both boxplots are constructed on the same scale, a comparison becomes easy. We can see that the Dvorak ratings are grouped considerable farther to the left, suggesting that the Dvorak keyboard is considerably easier to use. Also, the Qwerty ratings appear to have more spread because the Qwerty boxplot stretches farther to the right.

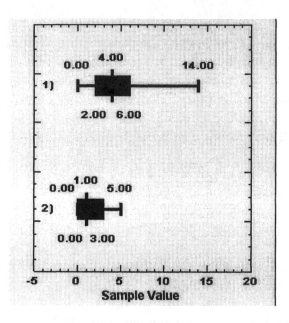

Important note: STATDISK generates boxplots based on the minimum, maximum, and three quartiles. STATDISK determines the values of the quartiles by following the same procedure described in Section 2-6 of *Elementary Statistics*, but other programs may use different procedures, so there may be some differences in boxplot results. The textbook states that there is not universal agreement on a single procedure for calculating quartiles, and different computer programs might yield different results. For example, if you use the data set of 1, 3, 6, 10, 15, 21, 28, and 36, you will get the results shown below. For this particular data set, STATDISK and the TI-83 Plus calculator agree, but they do not always agree.

	Q_1	Q_2	Q_3
STATDISK	4.5	12.5	24.5
Minitab	3.75	12.5	26.25
Excel	5.25	12.5	22.75
TI-83 Plus	4.5	12.5	24.5

2-4 Scatter Diagrams

Section 2-3 of *Elementary Statistics* describes the construction of scatter diagrams. A **scatter diagram** (or **scatterplot**) is a plot of the paired (x, y) data with a horizontal x axis and a vertical y axis. The textbook noted that the pattern of the plotted points is often helpful in determining whether there is some *relationship* between the two variables. Scatter diagrams are discussed at length in the Correlation and Regression (Chapter 9) chapter of the textbook.

Procedure for Generating a Scatter Diagram

To use STATDISK for generating a scatter diagram, you must have a collection of *paired* data.

1. Select **Data** from the main menu.

2. Select the subdirectory item of **Scatterplot**.

3. Enter or copy the paired data into columns 1 and 2. (You can, for example, retrieve a data set, then use Copy/Paste to move it into one of the columns in the Correlation and Regression module. Next, you can retrieve the second data set and use Copy/Paste to move it into the other column in the Correlation and Regression module.)

4. After the two matched columns of data have been entered, click on **Evaluate**.

The scatter diagram will include the straight line that fits the points best. This line is discussed in Chapter 9, but it can be ignored at this point in the course.

Section 2-3 of the textbook includes an example of a Minitab scatter diagram that uses the cigarette nicotine and tar data from Data Set 8 in Appendix B. If you use the same data with STATDISK, you will obtain the graph shown below. Based on the pattern of the points, we can conclude that there does appear to be a relationship between the amount of nicotine and the amount of tar in cigarettes. Cigarettes with greater amounts of nicotine appear to also have greater amounts of tar. Such relationships (or *correlations*) will be discussed at much greater length in Chapter 9.

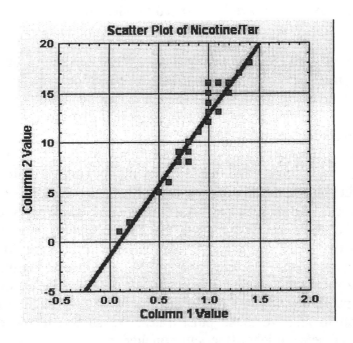

2-5 Sorting Data

On many different occasions, it becomes necessary to rearrange a data set so that the values are in order (ascending from low to high, or descending from high to low). First enter or copy the data into the Sample Editor module (accessed from the main menu item of Data), then click on the Format bar and proceed to select the way that you want the data arranged. Here are the details of this procedure.

Procedure for Sorting Data

1. Enter or retrieve a set of sample data. (To *enter* values, use **Data/Sample Editor** as described in Section 1-2 in this manual/workbook; to *retrieve* a data set, use **File/Open** as described in Section 1-4 of this manual/workbook.) After entering or retrieving a data set, you will be in the **Sample Editor** window.

2. With the data set listed in the Sample Editor window, click on the **Format** bar.

3. A window will appear with the title of Sample Format Options.

   ●Select a sort order of *ascending* if you want the data arranged in order from low to high.

   ●Select *descending* if you want the data arranged in decreasing order.

4. Click **OK** and the data will be arranged in the order you specified. The sorted data set can now be saved, or copied to other modules.

The sort feature is useful for identifying outliers. When analyzing data, it is important to identify outliers because they can have a dramatic effect on certain results. It is usually difficult to recognize an exceptional value when it is buried in the middle of a long list of values, but outliers become much easier to recognize with sorted data, because they will be found either at the beginning or end.

As an example, consider the pulse rates found in Data Set 14 in Appendix B of *Elementary Statistics*. We can retrieve that data set by clicking on **File**, then **Open**. We can scroll down to the file **pulse.sdd**. We then click **OK**. In the Sample Editor window, we can now click on **Format**. After selecting **Ascending** and clicking **OK**, we will get the STATDISK screen shown below. We can see that the pulse rates are arranged in order. We can see that the first two values of 8 and 15 are outliers in the sense that they are far away from all of the other pulse rates. Thinking about pulse rates, we recognize that humans cannot be alive if their pulse rates are 8 and 15. Because Data Set 14 is based on measurements from living statistics students, we now know that something is wrong. The outliers of 8 and 15 must be errors. If we plan to analyze those pulse rates, we should throw out these two wrong values, because their presence might affect our conclusions.

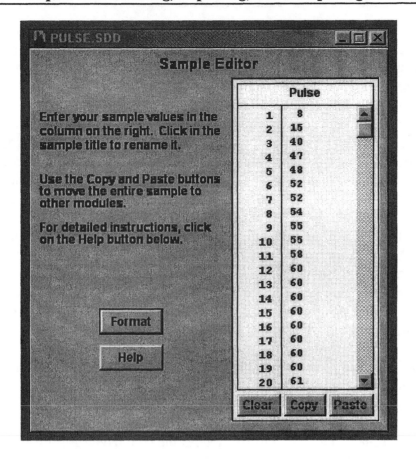

2-6 Generating Data Sets

In addition to entering or retrieving data, STATDISK allows you to generate several different types of data sets. Select the main menu item of Data, and the subdirectory includes these items:

Normal Generator
Uniform Generator
Binomial Generator
Poisson Generator
Coins Generator
Dice Generator
Frequency Table Generator

Some of these items will be discussed in the textbook after Chapter 2, but some of them are relevant now. The data generated by these modules can be saved or copied to other modules. The Normal Generator, Binomial Generator, and Poisson Generator are discussed in Chapters 4 and 5 of the textbook, but here are important comments about the other items. The frequency table generator is particularly useful for dealing with frequency tables discussed in Sections 2-4 and 2-5 of Chapter 2.

Frequency Table Generator: This feature can be used to obtain descriptive statistics for data that are summarized in the format of a frequency table. The basic approach is to generate a list of data values corresponding to a given frequency table. Click on **Data**, then **Frequency Table Generator** and proceed to enter the start and end values of the different classes. Click on **Generate** and you will get a window providing different options. If you want descriptive statistics for a frequency table, be sure to select the options of "Sample with Same Observed Freqs" and "Equal to Class Midpoints." After clicking **OK**, STATDISK will create a list of values that can be used in place of the frequency table. You can use Copy/Paste to copy the data to the Descriptive Statistics module, where you can find the mean, standard deviation, and other statistics.

Uniform Generator: You can generate numbers between a desired minimum value and maximum value. You can specify the number of decimal places (enter 0 if you want only whole numbers). The generated values are "uniform" in the sense that all of the generated values have the same chance of being selected. For example, if you select a sample size of 500, a minimum of 1, a maximum of 6, and 0 decimal places, the results simulate the rolling of a single die 500 times. (See also the Dice Generator described below.)

Coins Generator: You can select the number of coins to be tossed and the sample size (number of times those coins are tossed). The generated values are the numbers of heads that turn up.

Dice Generator: You can select the number of dice to be rolled, the sample size (number of times those dice are rolled), and the number of sides the dice have (use 6 for standard dice). The generated values are the totals of the dice.

We noted at the beginning of this chapter that the following are extremely important characteristics of data: center, variation, distribution, outliers, and pattern over time. Chapter 12 will consider the characteristic of pattern over time, but the other characteristics can all be investigated using tools of STATDISK, as we have described in this chapter. Here is a summary of the tools that are usually most relevant for the different characteristics:

1. **Center:** Use **Data/Descriptive Statistics** to find the mean and median.

2. **Variation:** Use **Data/Descriptive Statistics** to find the standard deviation, variance, and range.

3. **Distribution:** Use **Data/Histogram** and **Data/Boxplot** to generate a histogram and boxplot.

4. **Outliers:** Use **Data/Sample Editor/Format** to sort the data in ascending order, then examine the sample values to identify any that are very far away from almost all other values.

5. **Time:** See Chapter 12.

CHAPTER 2 EXPERIMENTS:
Describing, Exploring, and Comparing Data

2-1.   ***Comparing Regular Coke and Diet Coke***   In this experiment we use two small data sets as a quick introduction to using some of the basic STATDISK features. The data listed below are measured weights (in pounds) of random samples of the contents in cans of regular Coke and diet Coke.

      Regular Coke: 0.8192   0.8150   0.8163   0.8211   0.8181   0.8247
      Diet Coke: 0.7773   0.7758   0.7896   0.7868   0.7844   0.7861

   a. Find the indicated characteristics of the *regular* Coke sample data and enter the results below.

      Center: Mean: _____   Median: _____

      Variation: St. Dev.:_____   Variance:_____   Range:_____

      5-Number Summary: Min.:_____ Q_1:_____ Q_2:_____ Q_3:_____   Max.:_____

      Outliers: _____ Σx:_____ Σx^2:_____

   b. Find the indicated characteristics of the *diet* Coke sample data and enter the results below.

      Center: Mean: _____   Median: _____

      Variation: St. Dev.:_____   Variance:_____   Range:_____

      5-Number Summary: Min.:_____ Q_1:_____ Q_2:_____ Q_3:_____   Max.:_____

      Outliers: _____ Σx:_____ Σx^2:_____

   c. Compare the results from parts a and b. Also, try to provide an explanation for any notable differences.

      _____

      _____

      _____

2-2.   ***Effect of Adding a Constant***   Refer to the weights of *regular* Coke in Experiment 2-1. First add 2 to each value, then find the indicated characteristics and enter them in the spaces on the top of the following page. After adding 2 to each value, the weights are as follows:

2.8192   2.8150   2.8163   2.8211   2.8181   2.8247

Center: Mean: _____   Median: _____

Variation: St. Dev.:_____   Variance:_____   Range:_____

5-Number Summary: Min.:_____ Q_1:_____ Q_2:_____ Q_3:_____ Max.:_____

Outliers: _____ Σx:_____ Σx^2:_____

Based on a comparison of the results found here to those found in Experiment 2-1, answer the following questions.

How is the mean affected if the same constant is added to every value in a data set?

_____

How is the median affected if the same constant is added to every value in a data set?

_____

How is the standard deviation affected if the same constant is added to every value in a data set?

_____

If a new data set (different from the one used here or in Experiment 2-1) has mean $\bar{x} = 34.5$ and standard deviation $s = 6.7$, what is the mean and the standard deviation after 7 has been added to every value?

_____

2-3.   ***Effect of Multiplying by a Constant*** Refer to the weights of *regular* Coke in Experiment 2-1. First multiply each value by 10, then find the indicated characteristics and enter them in the spaces below. After multiplying each value by 10, the weights are as follows:

8.192   8.150   8.163   8.211   8.181   8.247

Center: Mean: _____   Median: _____

Variation: St. Dev.:_____   Variance:_____   Range:_____

5-Number Summary: Min.:_____ Q_1:_____ Q_2:_____ Q_3:_____ Max.:_____

Outliers: _____ Σx:_____ Σx^2:_____

Based on a comparison of the results found here to those found in Experiment 2-1, answer the following questions.

How is the mean affected if every value in a data set is multiplied by the same constant?

_____

_____

How is the median affected if every value in a data set is multiplied by the same constant?

_____

_____

How is the standard deviation affected if every value in a data set is multiplied by the same constant?

_____

_____

If a new data set (different from the one used here or in Experiment 2-1) has a mean of $\bar{x} = 34.5$ and $s = 6.7$, what is the mean and the standard deviation after every value has been multiplied by 5?

_____

_____

2-4.   *Retrieving a Data Set*   Experiments 2-1 through 2-3 could easily be done by manually entering sample data. If you refer to the CANS109 data set (Data Set 12 in Appendix B of *Elementary Statistics*), you will see 175 weights beginning with 270 lb. This data set is large enough so that manual entry of the data would likely result in typing errors. Instead of manual entry, we will simply recall the data set that already resides in the STATDISK program. To retrieve the file CANS109, click on **File**, click on **Open**, then scroll down the list of file names to select cans109.sdd, then click **OK**. The data should now be listed in the Sample Editor window. Now use Copy/Paste to copy the data to other modules so that you can find the following.

Center: Mean: _____   Median: _____

Variation: St. Dev.:_____   Variance:_____   Range:_____
                                              continued

5-Number Summary: Min.:_____ Q_1:_____ Q_2:_____ Q_3:_____ Max.:_____

Outliers: _____ Σx:_____ Σx^2:_____

Also obtain a printed copy of a histogram and boxplot. Based on the histogram, describe the shape of the distribution. Specifically, determine whether the distribution has a shape that is approximately bell-shaped.

_____

_____

2-5.   ***Retrieving a Data Set*** Repeat Experiment 2-4 using the CANS111 data. The STATDISK file name is cans111.sdd.

Center: Mean: _____   Median: _____

Variation: St. Dev.:_____   Variance:_____   Range:_____

5-Number Summary: Min.:_____ Q_1:_____ Q_2:_____ Q_3:_____ Max.:_____

Outliers: _____ Σx:_____ Σx^2:_____

Also obtain a printed copy of a histogram and boxplot. Based on the histogram, describe the shape of the distribution. Specifically, determine whether the distribution has a shape that is approximately bell-shaped.

_____

_____

Compare the results found here to those found in Experiment 2-3. (The CANS109 data set consists of axial loads for cans that are 0.0109 in. thick. An axial load of a can is measured by applying pressure on the top until the can collapses. The axial load is the weight required to make the can collapse.) The axial loads of the thinner cans are expected to be lower, indicating that those thinner cans are weaker. Do the sample statistics appear to support this expectation? When the cans have the top lids pressed into place, the top pressure is between 158 pounds and 165 pounds. Are the thinner cans acceptable?

_____

_____

_____

_____

2-6.   *Transforming a Data Set* In this experiment we investigate a process for transforming temperatures from the Fahrenheit scale to the Celsius scale. The following water temperatures (in degrees Fahrenheit) were recorded by an environmental scientist.

76 78 83 85 80 81 82 73 74 77   77 78

a. First use STATDISK to find the following.

   Center: Mean: _____   Median: _____

   Variation: St. Dev.:_____   Variance:_____   Range:_____

   5-Number Summary: Min.:_____ Q_1:_____ Q_2:_____ Q_3:_____ Max.:_____

   Outliers: _____ Σx:_____ Σx^2:_____

b. Now convert the temperatures from the Fahrenheit scale to the Celsius scale by using the formula $C = (5/9)(F - 32)$. To make that conversion, use Copy/Paste to copy the data from the Sample Editor module to column 1 in the Sample Transformations module, then use the equation $Y = (5/9)*(X1 - 32)$ to transform the data. (See Section 1-6.) Obtain the results for the transformed data and enter them below.

	Fahrenheit temperatures	Celsius temperatures
Mean	_____	_____
Median	_____	_____
St. Dev.	_____	_____
Q_1	_____	_____
Q_3	_____	_____

Compare the above results. What conclusions can be drawn?

_____

_____

_____

_____

2-7. ***Transforming a Data Set to z Scores*** Retrieve the CANS109 data and convert all the values to z scores. Accomplish this by using Copy/Paste to copy the data to the Sample Transformations module, where $z = (x -)/s$ is entered as follows. (The mean of the CANS109 data is 267.1142857, and the standard deviation is 22.112.)

$$Y = (X1 - 267.1142857)/22.112$$

Now find the mean and standard deviation of the z scores.

Mean:_____   Standard deviation:_____

Will these results be the same for *every* data set? _____

Now print a histogram and boxplot of the z scores. How is the shape of the distribution affected by the transformation to z scores?

_____

_____

2-8. ***Effect of Outlier*** In this experiment we will study the effect of an *outlier*. Use the same weights of *regular* Coke used in Experiment 2-1, but change the first entry from 0.8192 to 8192. (This type of mistake often occurs when the key for the decimal point is not pressed with enough force.) The outlier of 8192 is a mistake here, but outliers are sometimes correct values that differ substantially from the other sample values.

    Regular Coke: **8192** 0.8150 0.8163 0.8211 0.8181 0.8247

Using this modified data set that includes the outlier, find the following.

   Center: Mean: _____   Median: _____

   Variation: St. Dev.:_____   Variance:_____   Range:_____

   5-Number Summary: Min.:_____ Q_1:_____ Q_2:_____ Q_3:_____ Max.:_____

   Outliers: _____ Σx:_____ Σx^2:_____

Based on a comparison of these results to those found in Experiment 2-1, answer the following questions.

How is the mean affected by the presence of an outlier?

_____

_____

How is the median affected by the presence of an outlier?

_____

_____

How is the standard deviation affected by the presence of an outlier?

_____

_____

2-9.   *Exploring Distributions*   Retrieve each of the indicated data sets and obtain a printed copy of a histogram. Examine the histogram and described the general shape of the distribution. Determine whether the shape of the distribution is approximately bell-shaped.

Rainfall amounts for Boston on Tuesday (STATDISK file raintues.sdd):

_____

Weights of Domino sugar packets (STATDISK file sugar.sdd):

_____

Weights of quarters in circulation (STATDISK file quarters.sdd):

_____

Textbook prices for new books at the University of Massachusetts (STATDISK file newumass.sdd):

_____

Movie budget amounts (STATDISK file movbudg.sdd):

_____

2-10.   *Comparing Data Sets*   In some ways the following two data sets are very much alike but in other ways they are very different.

Data set A	83	84	81	78	77	76	75	87	87	87
Data set B	68	64	52	72	77	95	87	87	103	110

Use STATDISK to obtain graphs and descriptive statistics for each data set, compare the results, and then form your own conclusions about the similarities and differences. Justify your conclusions with references to the specific STATDISK results.

Similarities:

Differences:

2-11.   ***Comparing Ages of Oscar Winners***   Exercise 31 from Section 2-3 of the textbook lists the following ages of actors and actresses when they won Oscars.

Actors:	32	37	36	32	51	53	33	61	35	45	55	39
	76	37	42	40	32	60	38	56	48	48	40	43
	62	43	42	44	41	56	39	46	31	47	45	60
Actresses:	50	44	35	80	26	28	41	21	61	38	49	33
	74	30	33	41	31	35	41	42	37	26	34	34
	35	26	61	60	34	24	30	37	31	27	39	34

Using STATDISK, print the two boxplots for the two groups. Instead of two separate boxplots, display both boxplots in the same window, so that they both use the same scale. This makes comparisons much easier. Based on the results, compare the two groups and try to explain any difference.

_____

_____

_____

2-12.   ***Scatter Diagram***   Use STATDISK to print a scatter diagram for the tar and carbon monoxide data referred to in Exercise 23 from Section 2-3 of the textbook. The data can be copied from Data Set 8 in Appendix B of the textbook, or the data can be retrieved from files already stored on STATDISK. (The file names are tar.sdd and co.sdd.) Based on the scatter diagram, does there appear to be a relationship between cigarette tar and carbon monoxide? If so, describe the relationship.

_____

_____

2-13. *Scatter Diagram* Use STATDISK to print a scatter diagram for the distances around bear necks and the bear weights referred to in Exercise 24 from Section 2-3 of the textbook. The data can be copied from Data Set 7 in Appendix B of the textbook, or the data can be retrieved from the STATDISK files neck.sdd and weight.sdd. Based on the scatter diagram, does there appear to be a relationship between the distances around bear necks and the weights of the bears? If so, describe the relationship.

_____

_____

_____

2-14. *Working with Frequency Tables* Table 2-8 from the textbook is reproduced here. Use STATDISK to generate a list of sample values that can be used to represent the data in that table, then find the mean and standard deviation. (*Hint*: See "Frequency Table Generator" in Section 2-6 of this manual/workbook.) Enter the results here:

Mean:_____   Standard Deviation:_____

TABLE 2-8   Times (in minutes) Between
Old Faithful Eruptions

Time	Frequency
40-49	8
50-59	44
60-69	23
70-79	6
80-89	107
90-99	11
100-109	1

2-15. *Combining Data* Go to STATDISK's Sample Editor module and click on the Help bar. Carefully read the instructions for combining data sets into one big data set. Follow those directions and combine the different M&M data sets into one big data set. That is, retrieve and combine the files BLUE, BROWN, GREEN, ORANGE, RED, and YELLOW into one data set named M&M.

a. Save that data set and find the following results.

Center: Mean: _____   Median: _____

Variation: St. Dev.:_____   Variance:_____   Range:_____

5-Number Summary: Min.:_____ Q_1:_____ Q_2:_____ Q_3:_____ Max.:_____

Outliers: _____ Σx:_____ Σx^2:_____

b. Print a histogram of the data set.

c. Print a boxplot of the data set.

d. Describe the important characteristics of the data set. Be sure to address the nature of the distribution, measures of center, measures of variation, and any other important and notable features.

_____

_____

_____

2-16. ***Transformations of Data*** Earlier experiments involved the effects of transforming data by adding a constant to each value or by multiplying each value by a constant. Data sets are sometimes transformed by replacing each value with its square, or square root, or logarithm. Design and execute an experiment that reveals how the mean, median, and standard deviation are affected when each value x is replaced by the following expressions. (*Hint*: Use the **Sample Transformations** module.)

a. x^2_____

b. $\sqrt{x}$ _____

c. $\log x$_____

2-17. ***Working with Your Own Data*** Through observation or experimentation, collect your own set of values. Obtain at least 40 values and try to select data from an interesting population. Use STATDISK for help in answering the following questions.

a. Describe the nature of the data. That is, what do the values represent?

_____

_____

b. What method was used to collect the values?

_____

_____

c. What are some of the possible reasons why the data might not be representative of the true population? That is, what are some of the possible sources of bias?

_____

_____

d. Enter the appropriate values in the spaces provided and obtain printouts of a histogram and boxplot.

Center: Mean: _____   Median: _____

Variation: St. Dev.:_____   Variance:_____   Range:_____

5-Number Summary: Min.:_____ Q_1:_____ Q_2:_____ Q_3:_____ Max.:_____

Outliers: _____ Σx:_____ Σx^2:_____

e. Describe the shape of the distribution. Is the distribution approximately bell-shaped?

_____

_____

f. What particular characteristics of the data are noteworthy?

_____

_____

3

Probabilities Through Simulations

3-1 Simulation Methods

Chapter 3 in *Elementary Statistics* presents a variety of rules and methods for finding probabilities of different events. Except for Section 3-6, Chapter 3 of the textbook focuses on traditional approaches to computing probability values. This chapter, like Section 3-6 in the textbook, focuses instead on an alternative approach that is based on *simulations*. The textbook includes this definition:

Definition

A **simulation** of a procedure is a process that behaves the same way as the procedure, so that similar results are produced.

Mathematician Stanislaw Ulam once studied the problem of finding the probability of winning a game of solitaire, but the theoretical computations involved were too complicated. Instead, Ulam took the approach of programming a computer to simulate or "play" solitaire hundreds of times. The ratio of wins to total games played is the approximate probability he sought. This same type of reasoning was used to solve important neutron diffusion problems that arose during World War II. There was a need to determine how far neutrons would penetrate different materials, and the method of solution required that the computer make various random selections in much the same way that it can randomly select the outcome of the rolling of a pair of dice. This neutron diffusion project was named the Monte Carlo Project and we now refer to general methods of simulating experiments as *Monte Carlo methods*. Such methods are the focus of this chapter, where we illustrate how STATDISK can be used to simulate experiments so that probabilities can be estimated.

The concept of simulation is quite easy to understand with simple examples. (See the examples in *Elementary Statistics*, Section 3-6.)

- We could simulate the rolling of a die by using a computer to randomly generate whole numbers between 1 and 6 inclusive, provided that the computer selects from the numbers 1, 2, 3, 4, 5, and 6 in such a way that those outcomes are equally likely.

- We could simulate births by flipping a coin, where "heads" represents a baby girl and "tails" represents a baby boy. We could also simulate births by using a computer to randomly generate 1s (for baby girls) and 0s (for baby boys). Again, it is important to ensure that the 1s and 0s are equally likely, as in actual births.

The textbook includes a comment and an example that are so important that they are repeated here: *It is extremely important to construct a simulation so that it behaves just like the real procedure*. See this example and observe the right way and the wrong way of constructing the simulation.

EXAMPLE   Describe a procedure for simulating the rolling of a pair of dice.

SOLUTION   In the procedure of rolling a pair of dice, each of the two dice yields a number between 1 and 6 (inclusive) and those two numbers are then added. Any simulation should do the same thing.

Right way to simulate rolling two dice: Randomly generate one number between 1 and 6, then randomly generate another number between 1 and 6, then add the two results.

Wrong way to simulate rolling two dice: Randomly generate numbers between 2 and 12. This procedure is similar to rolling dice in the sense that the results are always between 2 and 12, but these outcomes between 2 and 12 are equally likely. With real dice, the values between 2 and 12 are *not* equally likely. This simulation would result in terrible results.

3-2   STATDISK Simulation Tools

STATDISK includes several different tools that can be used for simulations. If you click on the main menu item of **Data,** you get a submenu that includes these items:

> Normal Generator
> Uniform Generator
> Binomial Generator
> Poisson Generator
> Coins Generator
> Dice Generator
> Frequency Table Generator

Here are descriptions of these menu items:

- **Normal Generator**: Generates a sample of data randomly selected from a population having a normal (bell-shaped) distribution. You must enter the population mean and standard deviation.

- **Uniform Generator:** This tool is particularly good for the *random generation of integers*. It generates numbers between a desired minimum value and maximum value. You can specify the number of decimal places, so enter 0 if you want only whole numbers. The generated values are "uniform" in the sense that all possible values have the same chance of being selected. For example, if you select a sample size of 500, a minimum of 1, a maximum of 6, and 0 decimal places, the results simulate the rolling of a single die 500 times. (See also the Dice Generator described below.)

- **Binomial Generator**: Generates numbers of successes for a binomial probability distribution. You specify the number of values to be generated, the probability of success, and the number of trials in each case. Binomial probability distributions are not discussed until Chapter 4 in the textbook, so this item can be ignored for now.

- **Coins Generator**: This tool is particularly useful for those cases in which there are two possible outcomes (such as boy/girl) that are equally likely, as is the case with coin tosses. You select the number of generated values that you want, and you also select the number of coins to be tossed. The generated values are the numbers of heads that turn up.

- **Dice Generator:** You select the number of generated values that you want, and you select the number of dice to be rolled. You also select the number of sides the dice have (use 6 for standard dice). The generated values are the totals of the dice.

- **Frequency Table Generator:** This feature can be used to generate sample data drawn from a population that can be described by a frequency table. After entering the upper and lower class limits and the class frequencies, click on **Generate** to get the following dialog box. You can see that you have two basic options: (1) The generated data can have the same frequencies as those given in the table, or you can *randomly* generate sample values drawn from the population described by the frequency table; (2) each generated value can be set equal to a class midpoint, or its value can be randomly selected from the possible values in the class.

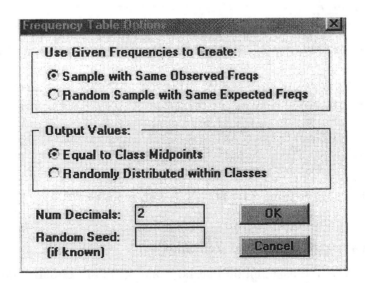

Random Seed

The above STATDISK tools include an option for entering a "random seed" if it is known. This entry will be usually left blank, but if you either record a seed that was used or if you enter a value for your own seed, you can duplicate results that were previously obtained. For example, an instructor might assign the generation of data with a particular random seed so that everyone in the class will get identical results. Most of the time you will not enter a value for the random seed so that your results will be different each time. This makes life a bit more interesting.

3-3 Sorting Data

In some cases, it is very helpful to *sort* data. (That is, arrange the data in order.) In Section 2-5 of this manual/workbook we described the procedure for sorting data. This procedure is slightly modified so that we begin with data that have been generated:

Procedure for Sorting Data

1. After generating sample data using the tools described in Section 3-2 of this manual/workbook, use Copy/Paste to copy the data to the **Sample Editor** module.

2. With the data set listed in the Sample Editor window, click on the **Format** bar.

3. A window will appear with the title of Sample Format Options.

   ●Select a sort order of *ascending* if you want the data arranged in order from low to high.

   ●Select *descending* if you want the data arranged in decreasing order.

4. Click **OK** and the data will be arranged in the order you specified. The "sorted" data set can now be saved, or copied to other modules.

3-4   Simulation Examples

We will now illustrate the preceding STATDISK features by describing specific simulations.

Simulation 1: Generating 50 births (boys/girls)

To simulate 50 births with the assumption that boys and girls are equally likely, use either of the following:

- Use STATDISK's **Uniform Generator** (see Section 3-2) to generate 50 integers between 0 and 1. Be sure to enter 0 for the number of decimal places. If you arrange the results in order, it is very easy to count the number of 0s (or boys) and the number of 1s (or girls). See Section 3-3 of this manual/workbook for the procedure to arrange data in order.

- Use STATDISK's **Coins Generator** (see Section 3-2). Enter 50 for the sample size and enter 1 for the number of coins. Again, it is very easy to count the number of 0s (or boys) and the number of 1s (or girls) if the data are sorted, as described in Section 3-3 of this manual/workbook.

Simulation 2: Rolling a single die 60 times

To simulate 60 rolls of a single die, use either of these approaches:

- Use STATDISK's **Uniform Generator** (see Section 3-2) to generate 60 integers between 1 and 6. (Enter 60 for the sample size and be sure to enter 0 for the number of decimal places.) Again, arranging them in order makes it easy to count the number of 1s, 2s, and so on.

- Use STATDISK's **Dice Generator**. Enter 60 for the sample size, enter 1 for the number of dice, and enter 6 for the number of sides.

Simulation 3: Generating 25 birth dates

Instead of generating 25 results such as "January 1," or "November 27," randomly generate 25 integers between 1 and 365 inclusive. (We are ignoring leap years). Use STATDISK's **Uniform Generator** and enter 25 for the sample size. Also enter 1 for the minimum, 365 for the maximum, and be sure to enter 0 for the number of decimal places (so that only integers are generated). If you sort the simulated birth dates by using the procedure in Section 3-3, it becomes easy to scan the sorted list and determine whether there are two birth dates that are the same. If there are two birth dates that are the same, they will show up as *consecutive* equal values in the sorted list.

CHAPTER 3 EXPERIMENTS: Probabilities through Simulations

3-1.   Use STATDISK's **Coins Generator** to simulate 500 trials, where each trial consists of tossing four coins and counting the number of heads. Sort the results, then find the number of cases in which exactly three heads turned up. Enter that value here:_____
Based on that result, estimate the probability of getting exactly three heads in four tosses of a coin. Enter the estimated probability here: _____

3-2.   Use STATDISK's **Dice Generator** to simulate 1000 rolls of a pair of dice. Sort the results, then find the number of times that the total was exactly 7. Enter that value here:_____   Based on that result, estimate the probability of getting a 7 when two dice are rolled. Enter the estimated probability here:_____   How does this estimated probability compare to the computed (theoretical) probability of 0.167?

_____

3-3.   Use STATDISK's **Dice Generator** to conduct a simulation that can be used to estimate the probability of getting a total of 10 when three dice are rolled. Enter the estimated probability here: _____   Describe the procedure used.

_____

3-4.   *Rolling 5 dice* Use STATDISK to conduct a simulation for estimating the probability of getting a total of 20 when 5 dice are rolled. Enter the estimated probability here:_____

3-5.   *Tossing 20 coins* Use STATDISK to conduct a simulation for estimating the probability of getting exactly 11 heads when 20 coins are tossed. Enter the estimated probability here:_____

3-6.   *Probability of exactly 11 girls* Use STATDISK to conduct a simulation for estimating the probability of getting exactly 11 girls when 20 babies are born. Enter the estimated probability here:_____
Describe the procedure used to obtain the estimated probability.

_____

3-7.   *Probability of at least 55 girls* Use STATDISK to conduct a simulation for estimating the probability of getting at least 55 girls in 100 births. Enter the estimated probability here:_____   Describe the procedure used to obtain the estimated probability.

_____

In testing a gender-selection method, assume that the Biogene Technology Corporation conducted an experiment with 100 couples who were treated, and that the 100 births included at least 55 girls. What should you conclude about the effectiveness of the treatment?

_____

3-8.   *Probability of at least 65 girls* Use STATDISK to conduct a simulation for estimating the probability of getting at least 65 girls in 100 births. Enter the estimated probability here:_____ Describe the procedure used to obtain the estimated probability.

In testing a gender-selection method, assume that the Biogene Technology Corporation conducted an experiment with 100 couples who were treated, and that the 100 births included at least 65 girls. What should you conclude about the effectiveness of the treatment?

3-9. **Roulette** Simulate the spinning of a roulette wheel 500 times by randomly generating 500 integers between 0 and 37. (American roulette wheels have slots numbered 0 through 36, plus another slot numbered 00. Consider a STATDISK-generated outcome of 37 to be 00 on the roulette wheel.) Arrange the results in ascending order. Assuming that you bet on the number 7 every time, how many times did you win? Based on the results, what is the estimated probability of winning if you bet on a single number?

Number of wins:_____ P(win) ≈ _____

In order to make a profit, your number of wins in 500 spins must be at least 14. Would you have made a profit?_____

3-10. **Birthdays** Simulate a class of 25 birth dates by randomly generating 25 integers between 1 and 365. (We will ignore leap years.) Arrange the birth dates in ascending order, then examine the list to determine whether at least two birth dates are the same. (This is easy to do, because any two equal integers must be next to each other.)

Generated "birth dates:"   ___ ___ ___ ___ ___ ___ ___ ___ ___ ___ ___ ___ ___

___ ___ ___ ___ ___ ___ ___ ___ ___ ___ ___ ___

Are at least two of the "birth dates" the same? _____

3-11. **Birthdays** Repeat Experiment 3-10 nine additional times and record all ten of the yes/no responses here: ___ ___ ___ ___ ___ ___ ___ ___ ___ ___

Based on these results, what is the probability of getting at least two birth dates that are the same (when a class of 25 students is randomly selected)? _____

3-12. **Birthdays** Repeat Experiments 3-10 and 3-11 for 50 people instead of 25. Based on the results, what is the estimated probability of getting at least two birth dates that are the same (when a class of 50 students is randomly selected)? _____

3-13. **Birthdays** Repeat Experiments 3-10 and 3-11 for 100 people instead of 25. Based on the results, what is the estimated probability of getting at least two birth dates that are the same (when a class of 100 students is randomly selected)? _____

3-14. **Normally distributed heights** Simulate 1000 heights of adult women. (Adult women have normally distributed heights with a mean of 63.6 in. and a standard deviation of 2.5 in.) Arrange the data in ascending order, then examine the results and estimate the probability of a randomly selected woman having a height between 64.5 in. and 72 in. (Those were the height restrictions for women to fit into Russian Soyuz spacecraft when NASA and Russia ran joint missions.) Enter the estimated probability here: _____

3-15. *Normally distributed bulb lives* The lifetimes of 75-watt light bulbs manufactured by the Lectrolyte Company have a mean of 1000 hours and a standard deviation of 100 hours. Generate a normally distributed sample of 500 bulb lifetimes by using the given mean and standard deviation. Examine the sorted results to estimate the probability of randomly selecting a light bulb that lasts between 850 hours and 1150 hours. Enter the result here._____

3-16. *Normally distributed IQ scores* IQ scores are normally distributed with a mean of 100 and a standard deviation of 15. Generate a normally distributed sample of 800 IQ scores by using the given mean and standard deviation. Sort the results (arrange them in ascending order).

   a. Examine the sorted results to estimate the probability of randomly selecting some-one with an IQ score between 90 and 110 inclusive. Enter the result here. _____

   b. Examine the sorted results to estimate the probability of randomly selecting some-one with an IQ score greater than 115. _____

   c. Examine the sorted results to estimate the probability of randomly selecting some-one with an IQ score less than 120. _____

   d. Repeat part a of this experiment nine more times and list all ten probabilities here.

   _____ _____ _____ _____ _____ _____ _____ _____ _____ _____

   e. Examine the ten probabilities obtained above and comment on the *consistency* of the results.

   _____

   f. How might we modify this experiment so that the results can become more consistent?

   _____

   g. If the results appear to be very consistent, what does that imply about any individ-ual sample result?

   _____

3-17. *Law of Large Numbers* In this experiment we test the Law of Large Numbers, which states that "as an experiment is repeated again and again, the empirical probability of success tends to approach the actual probability." (See Section 3-2 of the textbook.) We will use a simulation of a single die, and we will consider a success to be the outcome of a 1. (Based on the classical definition of probability, we know that $P(1) = 1/6 = 0.167$.)

   a. Simulate 5 trials by generating 5 integers between 1 and 6.
      Count the number of 6s that occurred and divide that number by 5 to get the
      empirical probability. Based on 5 trials, $P(1) \approx$ _____.
   b. Repeat part (a) for 25 trials. Based on 25 trials, $P(1) \approx$ _____.
   c. Repeat part (a) for 50 trials. Based on 50 trials, $P(1) \approx$ _____.
   d. Repeat part (a) for 500 trials. Based on 500 trials, $P(1) \approx$ _____.
   e. Repeat part (a) for 1000 trials. Based on 1000 trials, $P(1) \approx$ _____.
   f. In your own words, generalize these results in a restatement of the Law of Large

Numbers.

_____

_____

_____

3-18. *Sticky probability problem* Consider the following statement:

Two points along a straight stick are randomly selected. The stick is then broken at these two points. Find the probability that the three pieces can be arranged to form a triangle.

While a theoretical solution is possible, it is very difficult. Instead, we will use a computer simulation. The length of the stick is irrelevant, so assume it's one unit long and its length is measured from 0 at one end to 1 at the other end. Use STATDISK to randomly select the two break points with the random generation of two numbers from a uniform distribution with a minimum of 0, a maximum of 1, and 4 decimal places. Plot the break points on the "stick" below.

```
 ┌─────────────────────────────────────────┐
 0                                         1
```

A triangle can be formed if the longest segment is less than 0.5, so enter the lengths of the three pieces here: _____   _____   _____

Can a triangle be formed?   _____

Now repeat this process nine more times and summarize all of the results below.

Trial	Break Points		Triangle formed?
1			
2			
3			
4			
5			
6			
7			
8			
9			
10			

Based on the ten trials, what is the estimated probability that a triangle can be formed? _____ This estimate gets better with more trials.

4

Probability Distributions

4-1 Exploring Probability Distributions

Chapter 4 in *Elementary Statistics* introduces the important concept of a probability distribution, and only that chapter includes only *discrete* probability distributions. Section 4-2 introduces these important definitions:

Definitions

A **random variable** is a variable (typically represented by x) that has a single numerical value (determined by chance) for each outcome of some procedure.

A **probability distribution** gives the probability for each value of the random variable. (A probability distribution can be described various ways, including a table of x and $P(x)$ values, or a formula giving the probability for each value of x, or a graph depicting the values of x and their corresponding probabilities.)

When working with a probability distribution, we should consider the same important characteristics introduced in Chapter 2:

1. **Center:** Measure of center, which is a representative or average value that gives us an indication of where the middle of the data set is located

2. **Variation:** A measure of the amount that the scores vary among themselves

3. **Distribution:** The nature or shape of the distribution of the data, such as bell-shaped, uniform, or skewed

4. **Outliers:** Sample values that are very far away from the vast majority of the other sample values

5. **Time:** Changing characteristics of the data over time

Section 4-1 addresses these important characteristics for probability distributions. The characteristics of center and variation are addressed with formulas for finding the mean, standard deviation, and variance of a probability distribution. The characteristic of distribution is addressed through the graph of a probability histogram.

Although STATDISK is not designed to deal directly with a probability distribution, it can often be used. Let's consider Exercise 4 in Section 4-2. That exercise, reproduced below, requires that you first determine whether a probability distribution is defined by the given information and, if so, find the mean and standard deviation. If you examine the data in the table, you can verify that a probability distribution is defined because the two key requirements are satisfied: (1) The sum of the probabilities is equal to 1; (2) each individual probability is between 0 and 1 inclusive.

4. **TV Viewer Surveys** When four different households are surveyed on a Monday night, the random variable x is the number of households with televisions tuned to *Monday Night Football* on ABC (based on data from Nielsen Media Research). See the table on the next page.

x	$P(x)$
0	0.522
1	0.368
2	0.098
3	0.011
4	0.001

Having determined that the above table does define a probability distribution, let's now see how we can use STATDISK to find the mean and standard deviation. The basic approach is to use STATDISK's "Frequency Table Generator" to construct a table of actual values with the same distribution given in the table.

STATDISK Procedure for Working with a Probability Distribution
1.   Click on **Data**.
2.   Select the menu item of **Frequency Table Generator**.
3.   Enter class limits and frequencies that correspond to the probability distribution. Shown below are the entries corresponding to the probability distribution given in Exercise 4. See the first class where the value of 0 is represented by the class limits of -0.5 to 0.5 and a frequency of 522 (based on a probability of 0.522).

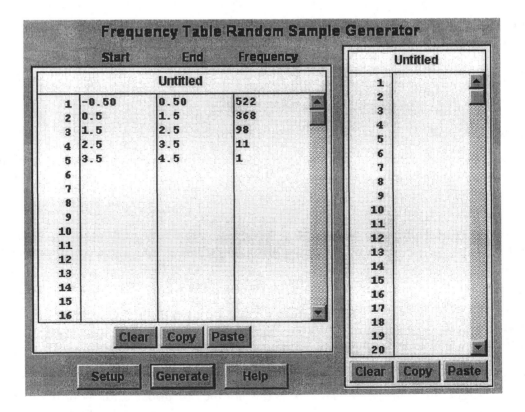

4. Click on the **Generate** bar, then select these options:
    - Sample with same observed freqs
    - Equal to class midpoints
5. Click **OK** and STATDISK will proceed to generate a set of data corresponding to the probability distribution.
6. You can now use Copy/Paste to copy the data to other modules where you can find the mean and standard deviation or you can construct a histogram. There's one correction: If you copy the data to the Descriptive Statistics module, the computed standard deviation and variance could be off a little, because the calculation assumes *sample* data, whereas we should consider the data to be a *population*. If the sample size is large, the discrepancies will be small. For the data from Exercise 4 reproduced above, the actual standard deviation is $\sigma = 0.71680$, but STATDISK yields $s = 0.71716$. The value of the mean will be correct. Here is a STATDISK histogram that shows the shape of the probability distribution:

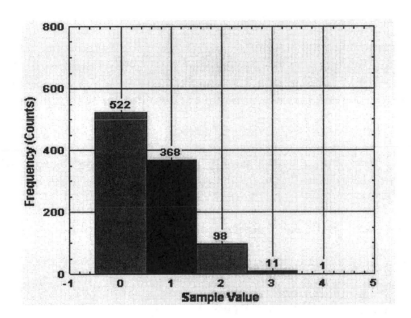

4-2 Binomial Distributions

In Section 4-3 of *Elementary Statistics* we define a **binomial distribution** to be a probability distribution that meets all of the following requirements:

1. The experiment must have a fixed number of trials.
2. The trials must be independent. (The outcome of any individual trial doesn't affect the probabilities in the other trials.)
3. Each trial must have all outcomes classified into two categories.
4. The probabilities must remain constant for each trial.

We also introduced notation with S and F denoting success and failure for the two possible categories of all outcomes. Also, p and q denote the probabilities of S and F, respectively, so that $P(S) = p$ and $P(F) = q$. We also use the following symbols.

   n denotes the fixed number of trials

   x denotes a specific number of successes in n trials so that x
   can be any whole number between 0 and n, inclusive.

   p denotes the probability of success in *one* of the n trials.

   q denotes the probability of failure in *one* of the n trials.

   $P(x)$ denotes the probability of getting exactly x successes among
   the n trials.

Section 4-3 describes three methods for determining probabilities in binomial experiments. Method 1 uses the binomial probability formula:

$$P(x) = \frac{n!}{(n - x)!x!} p^x q^{n-x}$$

Method 2 requires use of Table A-1, the table of binomial probabilities. Method 3 requires computer usage. We noted in Section 4-3 of the textbook that if a computer and software are available, this third method of finding binomial probabilities is fast and easy, as shown in the following STATDISK procedure.

STATDISK Procedure for Finding Probabilities with a Binomial Distribution

   1.   Click on **Analysis** from the main menu.

   2.   Select **Binomial Probabilities**.

   3.   You will now see a dialog box, so make these entries:

           -Enter the number of trials n.
           -Enter the probability of success p.

   4.   Click on **Evaluate**.

Section 4-3 of *Elementary Statistics* includes the following example:

Using Directory Assistance Use the binomial probability formula to find the probability of getting exactly 3 correct responses among 5 different requests from AT&T directory assistance. Assume that in general, AT&T is correct 90% of the time. That is, find $P(3)$ given that $n = 5$, $x = 3$, $p = 0.9$, and $q = 0.1$.

With STATDISK, click on **Analysis**, then select **Binomial Probabilities**, and proceed to enter 5 for the number of trials and 0.9 for the probability. Here is the STATDISK display:

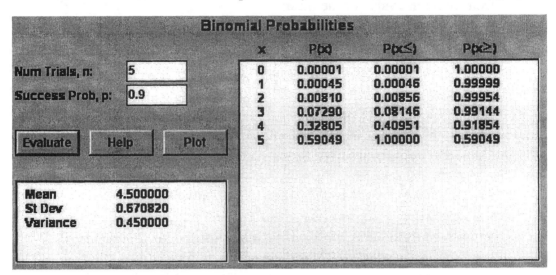

From this STATDISK display, we can see that $P(3) = 0.07290$. Note that the lower left corner of the display includes values of the mean, standard deviation, and variance. Also, STATDISK includes cumulative probabilities along with probabilities for the individual values of x. From the above display we can easily find probabilities such as these:

- The probability of 2 or fewer correct responses is 0.00856.
- The probability of 3 or more correct responses is 0.99144

The table of binomial probabilities (Table A-1) in Appendix B of the textbook includes limited values of n and p, but STATDISK is so much more flexible.

4-3 Poisson Distributions

See Section 4-5 of the textbook for a discussion of the Poisson distribution, where we see that it is a discrete probability distribution that applies to occurrences of some event *over a specified interval*. The random variable x is the number of occurrences of the event in an interval, such as time, distance, area, volume, or some similar unit. The probability of the event occurring x times over an interval is given by this formula:

$$P(x) = \frac{\mu^x \cdot e^{-\mu}}{x!} \qquad \text{where } e \approx 2.71828$$

We also noted that the Poisson distribution is sometimes used to approximate the binomial distribution when $n \geq 100$ and $np \leq 10$; in such cases, we use $\mu = np$. If using STATDISK, the Poisson approximation to the binomial distribution isn't used as often, because we can easily find binomial probabilities for a wide range of values for n and p.

STATDISK Procedure for Finding Probabilities for a Poisson Distribution

1.   Determine the value of the mean μ.

2.   Click on **Analysis** from the main menu.

3.   Select **Poisson Probabilities**.

4.   Enter the value of the mean μ.
      Click on the **Evaluate** button.

Section 4-5 of *Elementary Statistics* includes this example:

> **World War II Bombs** In analyzing hits by V-1 buzz bombs in
> World War II, South London was subdivided into 576 regions,
> each with an area of 0.25 km^2. A total of 535 bombs hit the
> combined area of 576 regions. If a region is randomly selected,
> find the probability that it was hit exactly twice.

Because a total of 535 bombs hit 576 regions, the mean number of hits is 535/576 = 0.929.
Having found the required mean μ, we can now proceed to use STATDISK. Click on
Analysis, select **Poisson Probabilities**, and enter 0.929 for the mean. The result will be as
shown below.

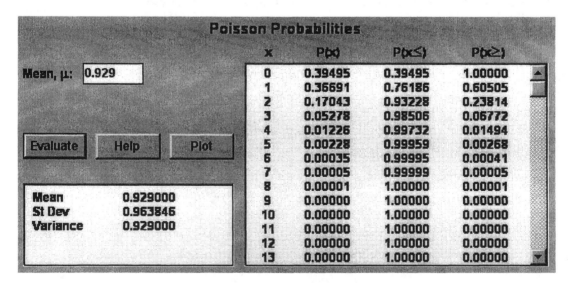

Note that the lower left corner includes values for the mean, standard deviation, and variance.
The probabilities and cumulative probabilities are listed in the right portion of the window.
For example, $P(2) = 0.17043$, which is the probability that a region would be hit two times.
The probability of a region being hit 0, 1, or 2 times is 0.93228. The probability of x having a
value of 2 or more is 0.23814, which is the probability of a region being hit at least twice.

4-4 Cumulative Probabilities

The main objective of this section is to reinforce the point that cumulative probabilities are often critically important. By *cumulative* probability, we mean the probability that the random variable x has a range of values instead of a single value. Here are typical examples:

- Find the probability of getting *at least* 13 girls in 14 births.
- Find the probability of getting *fewer than* 60 correct answers in 100 guesses to true/false questions.
- Find the probability of *more than* 5 wins when roulette is played 200 times.

Section 4-2 in *Elementary Statistics* makes the important point that in many cases, the cumulative probability is much more important than the probability of any individual event. As an example, the Chapter Problem for Chapter 4 notes that in a test of the MicroSort gender selection technique, there were 13 girls among 14 babies. Is this result unusual? Does this result really suggest that the technique is effective, or could it be that there were 13 girls among 14 babies just by chance? It was noted that in answering this key question, the relevant probability is the *cumulative* probability of getting 13 or more girls, not the probability of getting exactly 13 girls. The textbook supported that point with an example, reproduced here because it is so important:

Suppose you were flipping a coin to determine whether it favors heads, and suppose 1000 tosses resulted in 501 heads. This is not evidence that the coin favors heads, because it is very easy to get a result like 501 heads in 1000 tosses just by chance. Yet, the probability of getting exactly 501 heads in 1000 tosses is actually quite small: 0.0252. This low probability reflects the fact that with 1000 tosses, any specific number of heads will have a very low probability. However, we do not consider 501 heads among 1000 tosses to be unusual, because the probability of getting at least 501 heads is high: 0.488.

The textbook noted that the principle used in the above example can be generalized as follows:

Using Probabilities to Determine When Results are Unusual

- x successes among n trials is unusually *high* if $P(x$ or more$)$ is very small (such as less than 0.05)
- x successes among n trials is unusually *low* if $P(x$ or fewer$)$ is very small (such as less than 0.05)

Cumulative probabilities therefore play a critical role in identifying results that are considered to be *unusual*. Later chapters focus on this important concept.

CHAPTER 4 EXPERIMENTS: Probability Distributions

4-1.   Given below is Exercise 3 from Section 4-2 of *Elementary Statistics*. Use STATDISK with the procedure described in Section 4-1 of this manual/workbook for the following.

a.   Find the mean and standard deviation. Mean:_____ St. Dev.: _____

b.   The correct value of the standard deviation is $\sigma = 0.8660254038$. It was noted that the procedure described in Section 4-1 of this manual/workbook results in a standard deviation that is a little off. Compare the standard deviation to the correct value. _____

c.   Generate a histogram. Use a starting value of -0.5 and use a class width of 1. Obtain a printed copy of the result and describe the general shape of the distribution. _____

Gender Selection In a study of the MicroSort gender selection method, couples in a control group are not given a treatment and they each have three children. The probability distribution for the number of girls is given in the accompanying table.

x	P(x)
0	0.125
1	0.375
2	0.375
3	0.125

4-2.   Given below is Exercise 5 from Section 4-2 of *Elementary Statistics*. Use STATDISK with the procedure described in Section 4-1 of this manual/workbook for the following.

a.   Find the mean and standard deviation. Mean: _____ St. Dev.:_____

b.   The correct value of the standard deviation is $\sigma = 1$. It was noted that the procedure described in Section 4-1 of this manual/workbook results in a standard deviation that is a little off. Compare the standard deviation to the correct value. _____

c.   Generate a histogram. Use a starting value of -0.5 and use a class width of 1. Obtain a printed copy of the result and describe the general shape of the distribution. _____

Gender and Hiring If your college hires the next four employees without regard to gender, and the pool of applicants is large with an equal number of men and women, then the probability distribution for the number x of women hired is described in the accompanying table.

x	P(x)
0	0.0625
1	0.2500
2	0.3750
3	0.2500
4	0.0625

4-3. Given below is Exercise 6 from Section 4-2 of *Elementary Statistics*. Use STATDISK with the procedure described in Section 4-1 of this manual/workbook for the following.
 a. Find the mean and standard deviation. Mean: _____ St. Dev.:_____
 b. The correct value of the standard deviation is $\sigma = 1.253634716$ It was noted that the procedure described in Section 4-1 of this manual/workbook results in a standard deviation that is a little off. Compare the standard deviation to the correct value._____
 c. When finding the value of the standard deviation using the procedure from Section 4-1 of this manual/workbook, you could use frequencies of 4, 26, 36, ..., 2, or you could use frequencies of 40, 260, 360, ..., 20. Does the choice make a difference in the accuracy of the standard deviation? How? Why?

 _____

 _____

 d. Generate a histogram. Use a starting value of -0.5 and use a class width of 1. Obtain a printed copy of the result and describe the general shape of the distribution._____

Videotape Rentals *The accompanying table is constructed from data obtained in a study of the number of videotapes rented from Blockbuster.*

x	P(x)
0	0.04
1	0.26
2	0.36
3	0.20
4	0.08
5	0.04
6	0.02

4-4. Given below is Exercise 7 from Section 4-2 of *Elementary Statistics*. Use STATDISK with the procedure described in Section 4-1 of this manual/workbook for the following.
 a. Find the mean and standard deviation. Mean: _____ St. Dev.:_____
 b. The correct value of the standard deviation is $\sigma = 0.8894380248$. It was noted that the procedure described in Section 4-1 of this manual/workbook results in a standard deviation that is a little off. Compare the standard deviation to the correct value._____
 c. Generate a histogram. Use a starting value of -0.5 and use a class width of 1. Obtain a printed copy of the result and describe the general shape of the distribution._____

Prior Sentences *When randomly selecting a jail inmate convicted of DWI (driving while intoxicated), the probability distribution for the number x of prior DWI sentences is as described in the accompanying table (based on data from the U.S. Department of*

Justice).

x	P(x)
0	0.512
1	0.301
2	0.132
3	0.055

4-5.   ***Binomial Probabilities***   Use STATDISK to find the binomial probabilities corresponding to $n = 4$ and $p = 0.05$. Enter the results below, along with the corresponding results found in Table A-1 of the textbook.

x	P(x) from STATDISK	P(x) from Table A-1
0		
1		
2		
3		
4		

By comparing the above results, what obvious advantage does STATDISK have over Table A-1?

_____

4-6.   ***Binomial Probabilities***   Use STATDISK to find the binomial probabilities corresponding to $n = 4$ and $p = 1/4$ (or 0.25). Enter the results below.

x	P (x)
0	
1	
2	
3	
4	

If you attempt to use Table A-1 for these probabilities, you find that the table will not help. For this case, what obvious advantage does STATDISK have over Table A-1?

_____

4-7.   ***Binomial Probabilities***   Assume that boys and girls are equally likely and 100 births are randomly selected. Use STATDISK with $n = 100$ and $p = 0.5$ to find $P(x)$, where x represents the number of girls among the 100 babies.

a. $x = 35$ _____

b. $x = 45$ _____

c. $x = 50$ _____

4-8.   ***Binomial Probabilities***   In Experiment 4-7, we assumed that boys and girls are equally likely, but the actual values are $P(boy) = 0.5121$ and $P(girl) = 0.4879$. Repeat Experiment 4-3 using these values, then compare the results and write your conclusions in the space below and to the right.

a. $x = 35$ _____   _____

b. $x = 45$ _____   _____

c. $x = 50$ _____   _____

4-8. ***Binomial Probabilities*** In Experiment 4-3, we assumed that boys and girls are equally likely, but the actual values are P(boy) = 0.5121 and P(girl) = 0.4879. Repeat Experiment 4-3 using these values, then compare the results and write your conclusions in the space below and to the right.

a. $x = 35$ _____   _____

b. $x = 45$ _____   _____

c. $x = 50$ _____   _____

4-9. ***Cumulative Probabilities*** Assume that P(boy) = 0.5121, P(girl) = 0.4879, and that 100 births are randomly selected. Use STATDISK to find the probability that the number of girls among 100 babies is ...

a.   Fewer than 60 _____

b.   Fewer than 48 _____

c.   At most 30 _____

d.   At least 55 _____

e.   More than 40 _____

4-10. ***Identifying 0+*** In Table A-1 from the textbook, the probability corresponding to $n = 12$, $p = 0.10$, and $x = 6$ is shown as 0+. Use STATDISK to find the corresponding probability and enter the result here. _____

4-11. ***Identifying 0+*** In Table A-1 from the textbook, the probability corresponding to $n = 15$, p = 0.80, and $x = 5$ is shown as 0+. Use STATDISK to find the corresponding probability and enter the result here. _____

4-12. ***Identifying a Probability Distribution*** Use STATDISK to construct a table of x and $P(x)$ values corresponding to a binomial distribution in which $n = 26$ and $p = 0.3$. Enter the table in the margin.

In Experiments 4-13 through 4-16, use STATDISK to solve the exercise from Section 4-3 of Elementary Statistics.

4-13. **IRS Audits** The Hemingway Financial Company prepares tax returns for individuals. (Motto: "We also write great fiction.") According to the Internal Revenue Service, individuals making $25,000–$50,000 are audited at a rate of 1%. The Hemingway Company prepares 5 tax returns for individuals in that tax bracket, and 3 of them are audited.

a.   Find the probability that when 5 people making $25,000–$50,000 are randomly selected, exactly 3 of them are audited._____

b.   Find the probability that at least 3 are audited._____

c.   Based on the preceding results, what can you conclude about the Hemingway customers? Are they just unlucky, or are they being targeted for audits?

4-14. **On-Time Flights** The rates of on-time flights for commercial jets are continuously tracked by the U.S. Department of Transportation. Recently, Southwest Air had the best rate with 80% of its flights arriving on time. A test is conducted by randomly selecting 15 Southwest flights and observing whether they arrive on time.
  a.   Find the probability that exactly 10 flights arrive on time._____
  b.   Find the probability that at least 10 flights arrive on time._____
  c.   Find the probability that at least 10 flights arrive late._____
  d.   Would it be unusual for Southwest to have 5 flights arrive late? Why or why not?

4-15. **Directory Assistance** An article in *USA Today* stated that "Internal surveys paid for by directory assistance providers show that even the most accurate companies give out wrong numbers 15% of the time." Assume that you are testing such a provider by making 10 requests and also assume that the provider gives the wrong number 15% of the time.
  a.   Find the probability of getting 1 wrong number._____
  b.   Find the probability of getting at most 1 wrong number._____
  c.   If you do get at most 1 wrong number, does it appear that the rate of wrong numbers is not 15%?

4-16. **TV Viewer Surveys** The CBS television show *60 Minutes* has been successful for many years. That show recently had a share of 20, meaning that among the TV sets in use, 20% were tuned to *60 Minutes* (based on data from Nielsen Media Research). Assume that an advertiser wants to verify the 20% share value by conducting its own survey, and a pilot survey begins with 10 households having TV sets in use at the time of a *60 Minutes* broadcast.
  a.   Find the probability that none of the households are tuned to *60 Minutes*._____
  b.   Find the probability that at least one household is tuned to *60 Minutes*._____
  c.   Find the probability that at most one household is tuned to *60 Minutes*._____
  d.   If at most one household is tuned to *60 Minutes*, does it appear that the 20% share value is wrong? Why or why not?

In Experiments 4-17 through 4-20, use STATDISK to solve the given exercise from Section 4-5 of **Elementary Statistics.**

4-17. **Aircraft Hijackings** For the past few years, there has been a yearly average of 29 aircraft hijackings worldwide (based on data from the FAA). The mean number of hijackings per day is estimated as $\mu = 29/365$. If the United Nations is organizing a single international hijacking response team, there is a need to know about the chances of multiple hijackings in one day. Find the probability that the number of hijackings (x)

in one day is 0 or 1. _____
What do you conclude about the United Nation's organizing of a single response team?

4-18. **Deaths From Horse Kicks** A classic example of the Poisson distribution involves the number of deaths caused by horse kicks of men in the Prussian Army between 1875 and 1894. Data for 14 corps were combined for the 20-year period, and the 280 corps-years included a total of 196 deaths. After finding the mean number of deaths per corps-year, find the probability that a randomly selected corps-year has the following numbers of deaths.
a. 0 ____ **b.** 1 ____ **c.** 2 ____ **d.** 3 ____ **e.** 4 ____
The actual results consisted of these frequencies: 0 deaths (in 144 corps-years); 1 death (in 91 corps-years); 2 deaths (in 32 corps-years); 3 deaths (in 11 corps-years); 4 deaths (in 2 corps-years). Compare the actual results to those expected from the Poisson probabilities. Does the Poisson distribution serve as a good device for predicting the actual results?

4-19. **Homicide Deaths** In one year, there were 116 homicide deaths in Richmond, Virginia (based on "A Classroom Note On the Poisson Distribution: A Model for Homicidal Deaths In Richmond, VA for 1991," *Mathematics and Computer Education,* by Winston A. Richards). For a randomly selected day, find the probability that the number of homicide deaths is
a. 0 ____ **b.** 1 ____ **c.** 2 ____ **d.** 3 ____ **e.** 4 ____

Compare the calculated probabilities to these actual results: 268 days (no homicides); 79 days (1 homicide); 17 days (2 homicides); 1 day (3 homicides); there were no days with more than 3 homicides. _____

4-20. **Earthquakes** For a recent period of 100 years, there were 93 major earthquakes (at least 6.0 on the Richter scale) in the world (based on data from the *World Almanac and Book of Facts*). Assuming that the Poisson distribution is a suitable model, find the mean number of major earthquakes per year, then find the probability that the number of earthquakes in a randomly selected year is
a. 0 ____ **b.** 1 ____ **c.** 2 ____ **d.** 3 ____ **e.** 4 ____
f. 5 ____ **g.** 6 ____ **h.** 7 ____

Here are the actual results: 47 years (0 major earthquakes); 31 years (1 major earthquake); 5 years (3 major earthquakes); 2 years (4 major earthquakes); 0 years (5 major earthquakes); 1 year (6 major earthquakes); 1 year (7 major earthquakes). After comparing the calculated probabilities to the actual results, is the Poisson distribution a good model?

5

Normal Distributions

5-1 Generating Normal Data

There are many cases in which real data have a distribution that is approximately normal, and we can learn much about normal distributions by studying such cases. But how do we know when an "approximately" normal distribution is too far away from a perfect normal distribution? One way to circumvent this issue is to sample from a population with a known normal distribution and known parameters. Although this is often difficult to accomplish in reality, we can use the wonderful power of computers to obtain samples from theoretical normal distributions, and STATDISK has such a capability. Consider IQ scores. IQ tests are designed to produce a mean of 100 and a standard deviation of 15, and we expect that such scores are normally distributed. Suppose we want to learn about the variation of sample means for samples of IQ scores. Instead of going out into the world and randomly selecting groups of people, we can sample from theoretical populations. We can then learn much about the distribution of sample means. The following procedure allows you to obtain a random sample from a normally distributed population with a given mean and standard deviation.

STATDISK Procedure for Sampling from a Normally Distributed Population

1.   Select **Data** from the main menu at the top of the screen.

2.   Select **Normal Generator** from the subdirectory.

3.   You will now get a dialog box, so enter the following.

   -Enter the desired sample size (such as 500) for the number of values to be
    generated.
   -Enter the desired mean (such as 100).
   -Enter the desired standard deviation (such as 15).
   -Enter the desired number of decimal places for the generated values.
   -Enter a number for the Random Seed only if you want to *repeat* the generation
    of the data set. Otherwise, leave that box empty. (Leaving the Random Seed
    box empty causes a different data set to be randomly generated each time; if
    you use the same number as the seed, you will generate the same data set each
    time.)
   -Click on the **Generate** bar.

Shown on the top of the next page is the dialog box for generating 500 sample values (with 0 decimal places) from a normally distributed population with a mean of 100 and a standard deviation of 15. After clicking on **Generate**, the generated values will appear in the column at the right. You can then click on **Copy** so that the 500 sample values can be copied to other modules, such as the modules for creating a histogram, boxplot, or calculating the descriptive statistics.

The result of this process is a collection of *sample* data randomly generated from a population with the specified mean and standard deviation, so the mean of the sample data might not be exactly the same as the value specified, and the standard deviation of the sample data might not be exactly the same as the value specified. The sample generated in this case has a mean of 100.73 and a standard deviation of 15.416.

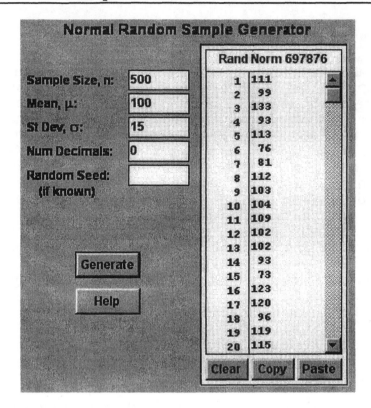

5-2 Normal Probabilities

Sections 5-2, 5-3, and 5-4 of *Elementary Statistics* describe methods for working with standard and nonstandard normal distributions. (Recall that a standard normal distribution has a mean of 0 and a standard deviation of 1.) Table A-2 in Appendix B of the textbook lists a wide variety of different *z* scores along with their corresponding areas. STATDISK can also be used to find probabilities associated with the normal distribution. STATDISK is much more flexible than the table, and it includes many more values. Here is the procedure.

STATDISK Procedure for Finding Probabilities with a Normal Distribution

1.   Select **Analysis** from the main menu at the top of the screen.

2.   Select **Probability Distributions** from the subdirectory.

3.   Select **Normal Distributions**.

4.   The mean and standard deviation have default values of 0 and 1 respectively, but they can be changed to any desired values. Press the **Enter** key, then click **OK** after changing those default values.

5.   Slide the mouse to the right and left to position it for the desired *z* or *x* value. (See the sample display on the top of the following page.) As the mouse moves, the vertical dashed line moves to the right and left, and the values in the box change accordingly.

6.   Zoom: When using the default mean and standard deviation of 0 and 1

respectively, you can get greater precision if you **Zoom** in by using the mouse to position the cursor just above and to the left of the region you want to enlarge. Hold down the left mouse button while dragging the mouse so that the rectangular box surrounds the region you want. When you release the mouse button, you will have zoomed in on a small portion of the curve, and you can get much better precision. For example, in the initial graph, you cannot position the mouse for $z = 1.645$, but you can get $z = 1.645$ by zooming in. After zooming in the first time, you can also zoom in again for even greater precision.

Using this procedure, we enter a mean of 100 and a standard deviation of 15 to obtain the display shown below. The position of the cursor indicates a value of $x = 115$, which is equivalent to $z = 1$. The following areas are indicated:

Below the curve and to the left of the cursor: 0.841345
Below the curve and to the right of the cursor: 0.158655
Twice the area in the tail bounded by the cursor: 0.317311
Twice the area below the curve and bounded by
    the centerline and the cursor: 0.682689
As Table A-2 (the area below the curve and
    bounded by the centerline and the cursor): 0.341345

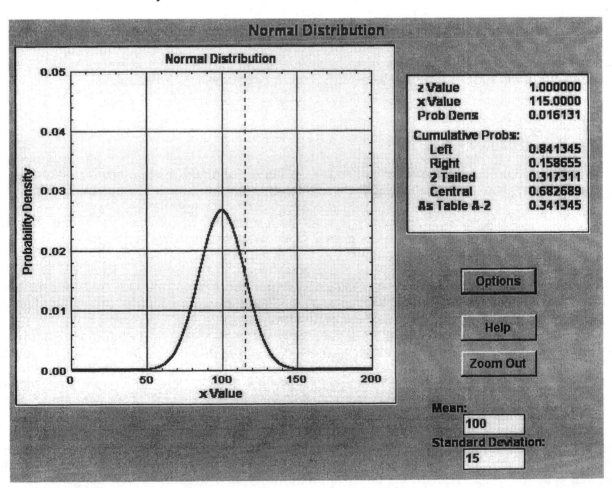

5-3 The Central Limit Theorem

Section 5-5 of the textbook introduces the central limit theorem, which is used throughout Chapters 6 and 7 of the textbook when the important topics of estimating parameters and hypothesis testing are discussed. From the textbook we have the following statement of the central limit theorem.

Central Limit Theorem

Given:
1.   The random variable x has a distribution (which may or may not be normal) with mean μ and standard deviation σ.
2.   Samples all of the same size n are selected from the population of x values in such a way that all samples of size n are equally likely.

Conclusions:
1.   The distribution of sample means $\bar{x}$ will, as the sample size increases, approach a *normal* distribution.
2.   The mean of the sample means will approach the population mean μ. (That is, the normal distribution from Conclusion 1 has mean μ.)
3.   The standard deviation of the sample means will approach $\sigma/\sqrt{n}$. (That is, the normal distribution from Conclusion 1 has standard deviation $\sigma/\sqrt{n}$).

In the study of methods of statistical analysis, it is extremely helpful to have a clear understanding of the statement of the central limit theorem. See Experiment 5-5.

5-4 Determining Normality

Section 5-7 of *Elementary Statistics* discusses criteria for determining whether sample data appear to come from a population having a normal distribution. These criteria are listed:

1.   **Histogram:** Construct a Histogram. Reject normality if the histogram departs dramatically from a bell shape. STATDISK can generate a histogram.
2.   **Outliers:** Identify outliers. Reject normality if there is more than one outlier present. (Just one outlier could be an error or the result of chance variation, but be careful, because even a single outlier can have a dramatic effect on results.) Using STATDISK, we can sort the data and easily identify any values that are far away from the majority of all other values.
3.   **Normal Quantile Plot:** If the histogram is basically symmetric and there is at most one outlier, construct a *normal quantile plot*. Examine the normal quantile plot and reject normality if the points do not lie close to a straight line, or if the points exhibit some systematic pattern that is not a straight-line pattern. STATDISK can generate a normal quantile plot. (Select **Data**, then **Normal Quantile Plot**.)

As an example, consider the STATDISK file BEARLEN, which consists of the lengths of a sample of bears that were anesthetized and measured. Shown below are the histogram and normal quantile plot. The histogram is approximately bell-shaped, which suggests that the lengths of bears have a normal distribution. The points in the normal quantile plot do lie close to a straight line, further suggesting that the lengths of bears are normally distributed. Also, we can determine that there are no outliers by copying the data to STATDISK's **Sample Editor**, where we can use **Format** to arrange the lengths in ascending order. Examining the sorted list shows that there are no exceptionally low or high values, so there are no outliers. Bottom line: The sample BEARLEN, which consists of lengths of bears, appears to come from a population having a normal distribution.

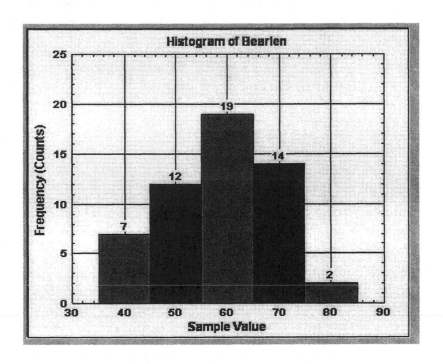

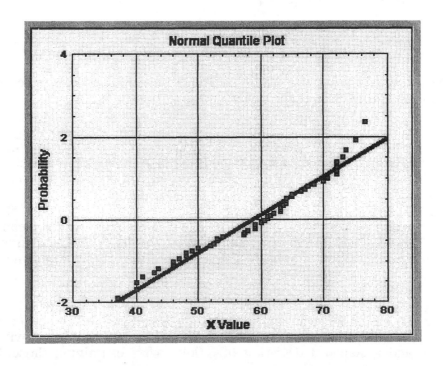

CHAPTER 5 EXPERIMENTS: Normal Distributions

5-1. **Identifying Significance** People generally believe that the mean body temperature is 98.6°F. Chapters 6 and 7 of *Elementary Statistics* include a sample of 106 body temperatures with these properties: The distribution is approximately normal, the sample mean is 98.20°F, and the standard deviation is 0.62°F. We want to determine whether these sample results differ from 98.6°F by a *significant* amount. One way to make that determination is to study the behavior of samples drawn from a population with a mean of 98.6.

     a.   Use STATDISK to generate 106 values from a normally distributed population with a mean of 98.6 and a standard deviation of 0.62. Use **Data/Descriptive Statistics** to find the mean of the generated sample. Record that mean here:____

     b.   Repeat part a nine more times and record the 10 sample means here:

         _____

     c.   By examining the 10 sample means in part b, we can get a sense for how much sample means vary for a normally distributed population with a mean of 98.6 and a standard deviation of 0.62. After examining those 10 sample means, what do you conclude about the likelihood of getting a sample mean of 98.20? Is 98.20 a sample mean that could easily occur by chance, or is it significantly different from the likely sample means that we expect from a population with a mean of 98.6?

         _____

     d.   Given that researchers did obtain a sample of 106 temperatures with a mean of 98.20°F, what does their result suggest about the common belief that the population mean is 98.6°F?

         _____

5-2. **Identifying Significance**   The U.S. Department of the Treasury claims that the procedure it uses to mint quarters yields a mean weight of 5.670 g.

     a.   Data Set 13 in Appendix B of the textbook lists the weights (in grams) of a sample of 50 quarters. That data set is stored as the STATDISK file QUARTERS. Open the file and find the sample mean and standard deviation.
Sample mean:_____   Standard deviation:_____

     b.   Generate 10 different samples, where each sample has 50 values randomly selected from a normally distributed population with a mean of 5.670 g and a standard deviation of 0.068 g (based on the U.S. Department of the Treasury specifications). For each sample, record the sample mean and enter it here.

         _____

     c.   By examining the 10 sample means in part b, we can get a sense for how much sample means vary for a normally distributed population with a mean of 5.670 and a standard deviation of 0.068. After examining those 10 sample means, what do you conclude about the likelihood of getting a sample mean of 5.670? Is 5.670 a sample mean that could easily occur by chance, or is it significantly different from the likely sample means that we expect from a population with a mean of 5.670?

_____

d.   Consider the sample mean found from Data Set 13 in Appendix B from the textbook. Does it suggest that the population mean of 5.670 g is not correct?

_____

5-3.   **Finding Probabilities for a Normal Distribution**   Use STATDISK's **Normal Distribution** module to find the indicated probabilities. First select **Analysis** from the main menu, then select **Probability Distributions**, then **Normal Distribution**.

a.   Given a population with a normal distribution, a mean of 0, and a standard deviation of 1, find the probability of a value greater than 1.50._____

b.   Given a population with a normal distribution, a mean of 100, and a standard deviation of 15, find the probability of a value less than 120._____

c.   Given a population with a normal distribution, a mean of 75, and a standard deviation of 10, find the probability of a value between 70 and 80._____

d.   Given a population with a normal distribution, a mean of 200, and a standard deviation of 20, find the probability of a value less than 190._____

e.   Given a population with a normal distribution, a mean of 200, and a standard deviation of 20, find the probability of a value between 190 and 250._____

5-4.   **Finding Values for a Normal Distribution** Use STATDISK's **Normal Distribution** module to find the indicated vales. First select **Analysis** from the main menu, then select **Probability Distributions**, then **Normal Distribution**.

a.   Given a population with a normal distribution, a mean of 0, and a standard deviation of 1, what value has an area of 0.000736 to its left?_____

b.   Given a population with a normal distribution, a mean of 100, and a standard deviation of 15, what value has an area of 0.000826 to its right?_____

c.   Given a population with a normal distribution, a mean of 75, and a standard deviation of 10, what value has an area of 0.999675 to its left?_____

d.   Given a population with a normal distribution, a mean of 200, and a standard deviation of 20, what value has an area of 0.000064 to its right?_____

e.   Given a population with a normal distribution, a mean of 200, and a standard deviation of 20, what value has an area of 0.000014 to its left?_____

5-5.   **Central Limit Theorem**   In this experiment, assume that all dice have 6 sides.

    a.   Use STATDISK to simulate the rolling of a single die 800 times. (Select **Data**, then **Dice Generator**.) Use Copy/Paste to copy the results to the Descriptive Statistics and Histogram modules, and enter the actual results below.

        One Die: Mean: _____

                      Standard Deviation: _____

                      Distribution shape: _____

    b.   Whereas part a used a single die, we will now use a pair of dice. Use STATDISK to "roll" two dice 800 times (again using Data/Dice Generator). The 800 values are *totals* for each pair of dice, so transform the totals to *means* by dividing each total by 2. (Use Copy/Paste to copy the results to the Sample Transformations module, then use the expression $Y = X1/2$ for the transformation.) Now use Copy/Paste to copy the 800 *means* to the Descriptive Statistics and Histogram modules. Enter the results below.

        Two Dice: Mean: _____

                      Standard Deviation: _____

                      Distribution shape: _____

    c.   Repeat part b using 10 dice. When finding the mean of the 10 dice, use the expression $Y = X1/10$ so that each total of the 10 dice is transformed to a mean.

        10 Dice: Mean: _____

                      Standard Deviation: _____

                      Distribution shape: _____

    d.   Repeat part b using 20 dice. When finding the mean of the 20 dice, use the expression $Y = X1/20$ so that each total of the 20 dice is transformed to a mean.

        20 Dice: Mean: _____

                      Standard Deviation _____

                      Distribution shape: _____

continued

e. General conclusions:

What happens to the mean as the sample size increases from 1 to 2 to 10 to 20?

_____

What happens to the standard deviation as the sample size increases?

_____

What happens to the distribution shape as the sample size increases?

_____

How do these results illustrate the central limit theorem?

_____

_____

_____

5-6.   **Determining Normality**   Refer to the indicated STATDISK data file. In each case, print a histogram, print a normal quantile plot, and identify any outliers. Based on the results, determine whether the sample data appear to come from a normally distributed population.

   a.   RAINFRI (amounts of rainfall in Boston on Fridays for a recent year)

   Outliers:_____   Normal Distribution?_____

   b.   CANS111 (axial loads of aluminum cans that are 0.0111 in. thick)

   Outliers:_____   Normal Distribution?_____

   c.   MCGWIRE (distances of homeruns hit by Mark McGwire in 1998)

   Outliers:_____   Normal Distribution?_____

   d.   HOMESELL (selling prices of homes in Dutchess County)

   Outliers:_____   Normal Distribution?_____

   e.   WATCH (errors of wristwatches in seconds)

   Outliers:_____   Normal Distribution?_____

6

Confidence Intervals and Sample Sizes

6-1 Confidence Intervals for Estimating μ

Section 6-2 of *Elementary Statistics* introduces confidence intervals as a means of estimating the value of a population parameter, such as the population mean μ. The following definition was included.

Definition
A **confidence interval** (or **interval estimate**) is a range (or an interval) of values used to estimate the true value of a population parameter.

Sections 6-2 and 6-3 from the textbook discuss the construction of confidence interval estimates of a population mean μ. The textbook stresses the importance of selecting the correct distribution (normal or t), but STATDISK automatically chooses the correct distribution based on the information that is entered. STATDISK is very easy to use for constructing confidence interval estimates for a population mean μ. However, STATDISK requires that you first obtain the descriptive statistics of n, $\bar{x}$, and s, as indicated in Step 1 of the following procedure.

STATDISK Procedure for Finding Confidence Intervals for μ

1. If the sample data are known but n, $\bar{x}$, and s are not yet known, find the values of those sample statistics by using STATDISK's Descriptive Statistics module. (See Section 2-1 of this manual/workbook.) You must know the values of n, $\bar{x}$, and s before proceeding to step 2.

2. Select **Analysis** from the main menu at the top of the screen.

3. Select **Confidence Intervals** from the subdirectory.

4. Select **Population Mean.**

5. You will now see a dialog box, so make these entries.

   -Enter a Confidence Level $(1 - \alpha)$, such as 0.95 or 0.99.
   -Enter the Sample Size n.
   -Enter the value of the sample mean $\bar{x}$.
   -Enter the value of the sample standard deviation s.
   -Enter the value of the population standard deviation σ if it is known. (The value of σ is usually unknown, so you will usually ignore this data-entry box.)

6. Click the **Evaluate** bar.

Section 6-2 of the textbook includes an example in which a 95% confidence interval is constructed for μ, given the following sample statistics obtained from the data in Table 6-1 of the textbook:

$$n = 106 \qquad \bar{x} = 98.20 \qquad s = 0.62$$

If you use these sample results with the above STATDISK procedure, the dialog box in Step 4 will appear as shown on the top of the next page. After clicking on the Evaluate bar, the results will be as shown on the right portion of the STATDISK display.

Confidence Interval for Population Mean

Confidence Level, (1-α):   `0.95`

Sample Size, n: `106`

Sample Mean, x̄:   `98.20`

Sample St Dev, s: `0.62`

Population St Dev, σ: ` `
(if known)

[Evaluate]   [Help]

Margin of error, E = 0.1180

95% confident that the population mean, μ, is within the range:

98.082 < μ < 98.318

Used Formula 6-1 with s for σ (see Help)

The textbook suggested that when using summary statistics, the confidence interval limits should be rounded to the same number of decimal places as those statistics, so we get this result for the 95% confidence interval:

$$98.08 < \mu < 98.32$$

After rounding, the STATDISK result is the same as the textbook result.

6-2 Confidence Intervals for Estimating p

The STATDISK procedure for finding confidence interval estimates of a population proportion p is very similar to the procedure for finding confidence interval estimates of μ. The program requires the sample size n and the number of successes x. In some cases, the values of x and n are both known, but in other cases the given information may consist of n and a sample percentage. For example, Exercise 23 in Section 6-5 provides the information that "pilots died in 5.2% of 8411 crash landings (based on data from 'Risk Factors for Pilot Fatalities in General Aviation Airplane Crash Landings,' by Rostykus, Cummings, and Mueller, *Journal of the American Medical Association*, Vol. 280, No. 11)." Based on that information, we know that $n = 8411$ and $\hat{p} = 0.052$, but we do not yet know the number of successes x that is needed for STATDISK. Because $\hat{p} = x/n$, it follows that $x = \hat{p}n$, so the number of successes can be found by multiplying the sample proportion $\hat{p}$ and the sample size n. Given that $n = 8411$ and $\hat{p} = 0.052$, we calculate that $x = (0.052)(8411) = 437.372$. We must round this result, because x must be a whole number. We get $x = 437$.

To find the number of successes x from the sample proportion and sample size:

Calculate $x = \hat{p}n$ and round the result to the nearest whole number.

After having determined the value of the sample size n and the number of successes x, we can proceed to use STATDISK as follows.

STATDISK Procedure for Finding Confidence Intervals for *p*

1. Select **Analysis** from the main menu.

2. Select **Confidence Intervals**.

3. Select **Population Proportion**.

4. Make these entries in the dialog box:

    -Enter a confidence level, such as 0.95 or 0.99.
    -Enter the value for the sample size *n*.
    -Enter the number of successes for *x*.

5. Click on the **Evaluate** bar.

Shown below is the STATDISK display for finding the 95% confidence interval described in Exercise 23 of Section 6-5.

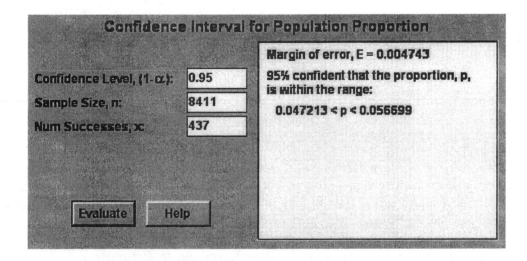

6-3 Confidence Intervals for Estimating σ

The construction of confidence interval estimates of a population standard deviation σ or variance σ^2 is described in Section 6-6 of the textbook. After selecting a confidence level and entering the sample size *n* and sample standard deviation *s*, STATDISK will automatically provide a confidence interval estimate of σ along with a confidence interval estimate of σ^2. You get both confidence intervals (for σ and σ^2) whether you want them or not. Be careful when correctly identifying the value of the sample standard deviation *s*. Be careful to enter the sample standard deviation where it is required; if only the sample variance is known, find its square root and enter that value for *s*. After obtaining the values of the sample size *n* and sample standard deviation *s*, proceed as follows.

STATDISK Procedure for Finding Confidence Intervals for σ and σ^2

1. Select **Analysis** from the main menu.

2. Select **Confidence Intervals**.

3. Select **Population St Dev**.

4. Make these entries in the dialog box:

> -Enter a confidence level, such as 0.95 or 0.99.
> -Enter the value for the sample size n.
> -Enter the value of the sample *standard deviation s* (not the *variance s^2*)

5. Click on the **Evaluate** bar.

6-4 Sample Sizes for Estimating μ

Section 6-4 of the textbook discusses the procedure for determining the sample size necessary to estimate a population mean μ. We use the formula

$$ n = \left[\frac{z_{\alpha/2}\,\sigma}{E} \right]^2 $$

STATDISK requires that we know the desired degree of confidence, the margin of error E, and the population standard deviation σ. The textbook notes that it is unusual to know σ without knowing μ, but σ might be known from a previous study or it might be estimated from a pilot study or the range rule of thumb. The entry of a finite population size N is optional, as described in the following steps.

STATDISK Procedure for Finding Sample Sizes Required to Estimate μ

1. Select **Analysis** from the main menu.

2. Select **Sample Size Determination** from the subdirectory.

3. Select the option of **Estimate Mean**.

4. In the dialog box, make these entries:

> -Enter a confidence level, such as 0.95 or 0.99.
>
> -Enter the value of the population standard deviation σ. (If σ is not known, consider estimating it from a previous study or pilot study or use the range rule of thumb.)
> -For the entry box labeled Population Size N: Leave it blank if you are sampling with replacement, or if you have a small sample drawn from a large population.

(Consider a sample size n to be "small" if $n \leq 0.05N$.) Enter a value only if you are sampling without replacement from a finite population with known size N, and the sample is large so that $n > 0.05N$. This box is usually left blank.

5. Click on the **Evaluate** bar.

Here is the example from Section 6-4. It is followed by the STATDISK display.

IQ Scores of Statistics Professors Assume that we want to estimate the mean IQ score for the population of statistics professors. How many statistics professors must be randomly selected for IQ tests if we want to be 95% confidence that the sample mean is within 2 IQ points of the population mean?

Sample Size Required to Estimate Mean

Required sample size is:

Confidence Level, (1-α): `0.95`

$n = 217$

Margin of Error, E: `2`

Assumed either infinite population or the population was sampled with replacement

Population St Dev, σ: `15`

Population Size, N:
(if known)

[Evaluate]   [Help]

6-5 Sample Sizes for Estimating p

Section 6-5 of the textbook describes methods for determining the sample size needed to estimate a population proportion p. STATDISK requires that you enter a confidence level (such as 0.95) and a margin of error E (such as 0.03). In addition to those two required entries, there are also two optional entries. You can enter an estimate of p if one is known, based on such factors as prior knowledge or results from a previous study. You can also enter the population size N if it is known and if you are sampling without replacement. The textbook includes the following two cases.

When an Estimate $\hat{p}$ Is Known: Formula 6-5 $n = \dfrac{\left[z_{\alpha/2}\right]^2 \hat{p}\hat{q}}{E^2}$

When No Estimate $\hat{p}$ Is Known: Formula 6-6 $n = \dfrac{\left[z_{\alpha/2}\right]^2 \cdot 0.25}{E^2}$

STATDISK Procedure for Finding Sample Sizes Required to Estimate p

1. Select **Analysis** from the main menu.

2. Select the subdirectory item of **Sample Size Determination**.

3. Select **Estimate Proportion**.

4. Make these entries in the dialog box:

   -Enter a confidence level, such as 0.95 or 0.99.
   -Enter a margin of error E. (The margin of error must be expressed in decimal form. For example, a margin of error of "three percentage points" should be entered as 0.03.)
   -Enter an estimated proportion if it is known. (This value might come from a previous study, or from knowledge about the value of the sample proportion. If such a value is not known, leave this box empty.)
   -Enter a value for the population size N if you will sample without replacement from a finite population of N subjects. (Because the population size is typically large or because sampling is done with replacement, this box is usually left blank.)

5. Click the **Evaluate** bar.

6-6 Sample Sizes for Estimating σ

Section 6-6 of *Elementary Statistics* describes a procedure for determining the sample size required to estimate a population standard deviation σ or population variance σ^2. Table 6-2 in the textbook lists sample sizes for several different cases, but STATDISK is much more flexible and allows you to find the sample size for many other cases.

STATDISK Procedure for Finding Samples Sizes Required to Estimate σ or σ^2

1. Select **Analysis** from the main menu.

2. Select **Sample Size Determination**.

3. Select the third option of **Estimate St Dev**.

4. Make these entries in the dialog box:

   -Enter a confidence level, such as 0.95 or 0.99.
   -Enter a "Percent Margin of Error, %E." (For example, if you enter 20, STATDISK will provide the sample size required so that s is within 20% of σ, and it will also provide the sample size required so that s^2 is within 20% of σ^2.

5. Click on the **Evaluate** bar.

Statistics textbooks rarely address the issue of determining sample size for estimating a population standard deviation σ or variance σ^2, but *Elementary Statistics* discusses this topic in Section 6-6 and STATDISK now allows you to do calculations with ease.

CHAPTER 6 EXPERIMENTS: Confidence Intervals and Sample Sizes

6-1. **Confidence Interval for Estimating a Mean**   Exercise 21 in *Elementary Statistics* includes reference to a sample of weights of the cola in cans of regular Coke. Those weights are listed below, but they are also stored as a file (ckregwt.sdd) in STATDISK, so it is not necessary to enter these values. Instead, retrieve the file and proceed to find the values of n, $\bar{x}$, and s. You can refer to Section 2-1 of this manual workbook, but here is a summary: Click on **File**, then **Open**, then scroll to **ckregwt.sdd** and double click on that file name. Click on **Copy**, then **Data**, **Descriptive Statistics**, **Paste**, then **Evaluate**. Record the values of the sample statistics and enter them here.

$n = \underline{\hspace{2cm}}$ $\bar{x} = \underline{\hspace{2cm}}$ $s = \underline{\hspace{2cm}}$

Weights (in pounds) of a sample of cans of regular Coke

0.8192	0.8150	0.8163	0.8211	0.8181	0.8247
0.8062	0.8128	0.8172	0.8110	0.8251	0.8264
0.7901	0.8244	0.8073	0.8079	0.8044	0.8170
0.8161	0.8194	0.8189	0.8194	0.8176	0.8284
0.8165	0.8143	0.8229	0.8150	0.8152	0.8244
0.8207	0.8152	0.8126	0.8295	0.8161	0.8192

Use STATDISK to find the following confidence interval estimates of the population mean μ.

99.5% confidence interval: $\underline{\hspace{6cm}}$

99% confidence interval: $\underline{\hspace{6cm}}$

98% confidence interval: $\underline{\hspace{6cm}}$

95% confidence interval: $\underline{\hspace{6cm}}$

90% confidence interval: $\underline{\hspace{6cm}}$

After examining the pattern of the above confidence intervals, complete the following. "As the degree of confidence decreases, the confidence interval limits

$\underline{\hspace{10cm}}$."

In your own words, explain why the preceding completed statement makes sense. That is, why should the confidence intervals behave as you have described.

$\underline{\hspace{12cm}}$

$\underline{\hspace{12cm}}$

6-2. **Effect of an Outlier**   Using the same sample data from Experiment 6-1, change the first weight from 0.8192 lb to 8192 lb (a common error in data entry) and find the confidence intervals for the population mean. Enter the results on the top of the following page.

99.5% confidence interval: _____

99% confidence interval: _____

98% confidence interval: _____

95% confidence interval: _____

90% confidence interval: _____

By comparing these results to the results from Experiment 6-1, what do you conclude about the effect of an outlier on the values of the confidence interval limits?

_____

_____

The outlier introduced in this experiment is an error, because it involves the deletion of a decimal point that should be present. Are all outliers errors? Explain.

_____

_____

6-3.   **Effect of Multiplying by a Constant** Repeat Experiment 6-1 after multiplying each of the sample scores by 10. Enter the resulting confidence intervals below.

99.5% confidence interval: _____

99% confidence interval: _____

98% confidence interval: _____

95% confidence interval: _____

90% confidence interval: _____

After comparing these results to those obtained in Experiment 6-1, what do you conclude about the effect of multiplying each sample value by the same constant?

_____

_____

The original sample weights in Experiment 6-1 are in *pounds*. Assume that you want to obtain a confidence interval based on *ounces*. (1 pound = 16 ounces) One way to obtain such a confidence interval is to multiply each of the original sample values by 16, then proceed to obtain the confidence interval. However, there is a much easier way to obtain a confidence interval based on ounces. First enter the 95% confidence interval based on ounces, then on the top of the next page, describe the method used to obtain that result.

95% confidence interval: _____

Method:_____

_____

_____

6-4. **Comparing Diet Coke and Regular Coke** Experiment 6-1 is based on sample data consisting of weights of the cola in cans of *regular* Coke. Listed below are weights of the cola in a sample of cans of *diet* Coke. Those values are already stored in STATDISK under the file name of ckdietwt.sdd, so they can be retrieved and it is not necessary to manually enter them through a process that can best be described as tedious.

Weights (in pounds) of a sample of cans of diet Coke

0.7773	0.7758	0.7896	0.7868	0.7844	0.7861
0.7806	0.7830	0.7852	0.7879	0.7881	0.7826
0.7923	0.7852	0.7872	0.7813	0.7885	0.7760
0.7822	0.7874	0.7822	0.7839	0.7802	0.7892
0.7874	0.7907	0.7771	0.7870	0.7833	0.7822
0.7837	0.7910	0.7879	0.7923	0.7859	0.7811

Record the values of the sample statistics and enter them here.

$n =$ _____ $\bar{x} =$ _____ $s =$ _____

Use STATDISK to find the following confidence interval estimates of the population mean μ.

99.5% confidence interval: _____

99% confidence interval: _____

98% confidence interval: _____

95% confidence interval: _____

90% confidence interval: _____

Based on comparisons of the sample statistics and confidence intervals listed here to those listed in Experiment 6-1, what do you conclude about the difference between the mean weight of cola in cans of *regular* Coke and the mean weight of cola in cans of *diet* Coke? Does there appear to be a significant difference? What is the basis for your conclusion?

_____

_____

If there does appear to be a significant difference, what might be a likely explanation for it? That is, why would regular Coke weigh more (or less) than diet Coke?

_____

_____

6-5.   **Skull Breadths** Exercise 13 in Section 6-2 of *Elementary Statistics* refers to a sample of 35 skulls obtained for Egyptian males who lived around 1850 B.C. The maximum breadth of each skull is measured with the result that $\bar{x} = 134.5$ mm and $s = 3.48$ mm (based on data from *Ancient Races of the Thebaid* by Thomson and Randall-Maciver). Use STATDISK with these sample results to construct a 95% confidence interval for the mean maximum breadth for all such skulls. Also, write a statement that interprets the confidence interval.

_____

_____

6-6.   **Second-Hand Smoke**   Exercise 24 in Section 6-2 of *Elementary Statistics* refers to sample data consisting of the measured serum cotinine levels in three different groups of people. Serum cotinine is a metabolite of nicotine, meaning that cotinine is produced when nicotine is absorbed by the body. Higher levels of cotinine correspond to higher levels of exposure to smoke that contains nicotine. STATDISK includes the files containing the cotinine levels for the three groups of people. The STADISK file NOETS lists the cotinine levels for subjects who are nonsmokers and have no exposure to environmental tobacco smoke at home or work. The STATDISK file ETS lists cotinine levels for subjects who are nonsmokers exposed to tobacco smoke at home or work. The STATDISK file SMOKERS lists cotinine levels for subjects who report tobacco use. For each of the three groups, construct confidence interval estimates of the population mean of all people in the group, them compare the confidence intervals. Based on the results, does there appear to be enough evidence to conclude that exposure to tobacco smoke corresponds to higher levels of cotinine? What does that conclusion imply about the effects of second-hand smoke?

NOETS confidence interval: _____

ETS confidence interval: _____

SMOKERS confidence interval: _____

_____

_____

6-7.   **Axial Loads of Cola Cans**   Data Set 12 in Appendix B of *Elementary Statistics* includes two sets of data. The first set consists of axial loads for a sample of 175 cola cans that are 0.0109 in. thick, and the second data set consists of axial loads for a sample of 175 cola cans that are 0.0111 in. thick. The axial load is measured by applying pressure to the top of the can until it collapses. It is important to have axial loads high enough to withstand the pressure that is applied when the top lid is pressed into place. Instead of manually entering the total of 350 sample values, use STATDISK to retrieve the files cans109.sdd and cans111.sdd. Proceed to find confidence intervals.

0.0109 in. cans:_____

0.0111 in. cans:_____

Compare the above results. Are the 0.0109 in. cans significantly weaker?

_____

6-8. **Car Pollution** Exercise 17 in Section 6-3 of *Elementary Statistics* refers to a sample of 7 cars. Each car was tested for nitrogen oxide emissions (in grams per mile) and the following results were obtained: 0.06, 0.11, 0.16, 0.15, 0.14, 0.08, 0.15 (based on data from the Environmental Protection Agency). Assuming that this sample is representative of the cars in use, use STATDISK to construct a 95% confidence interval estimate of the mean amount of nitrogen oxide emissions for all cars. Enter the result here.

Unlike Experiments 6-1 through 6-7, this experiment involves a *small* sample. The procedures for dealing with small samples are described in Section 6-3 of the textbook. Does STATDISK require any special handling to account for the fact that we are dealing with a small sample? Are the STATDISK results consistent with the procedure described in the textbook?

6-9. **Simulated Data** STATDISK is designed to generate random numbers from a variety of different sampling distributions. In this experiment we will generate 500 IQ scores, then we will construct a confidence interval based on the sample results. IQ scores have a normal distribution with a mean of 100 and a standard deviation of 15. First generate the 500 sample values as follows.

   1. Click on **Data,** then select **Normal Generator**.

   2. In the dialog box, enter a sample size of 500, a mean of 100, and a standard deviation of 15. For the number of decimal places, enter 0. Click on the **Evaluate** bar.

   3. Use Copy/Paste to copy the 500 generated sample values and paste them in the Descriptive Statistics module so that the sample mean and sample standard deviation can be found. Record these results here.

   $n =$ _____ $\bar{x} =$ _____ $s =$ _____

Now use the sample statistics to construct a 95% confidence interval estimate of the population mean of all IQ scores. Enter the 95% confidence interval here.

Because of the way that the sample data were generated, we *know* that the population mean is 100. Do the confidence interval limits contain the true mean IQ score of 100?

If this experiment were to be repeated over and over again, how often would we expect the confidence interval limits to contain the true population mean value of 100? Explain how you arrived at your answer.

6-10. **Simulated Data** Follow the same three steps listed in Experiment 6-9 to randomly generate 500 IQ scores from a population having a normal distribution, a mean of 100, and a standard deviation of 15. Record the sample statistics here.

$n = $ _____ $\bar{x} = $ _____ $s = $ _____

Confidence intervals are typically constructed with confidence levels around 90%, 95%, or 99%. Instead of constructing such a typical confidence interval, use the above sample statistics with STATDISK to construct a 50% confidence interval. List the result here.

_____

Does the above confidence interval have limits that actually do contain the true population mean, which we know is 100? _____

Repeat the above procedure 9 more times and list the resulting 50% confidence intervals here.

_____   _____   _____   _____   _____

_____   _____   _____   _____

Among the total of the 10 confidence intervals constructed, how many of them actually do contain the true population mean of 100? Is this result consistent with the fact that the level of confidence used is 50%? Explain.

_____

_____

6-11. **Combining Data Sets** Refer to Data Set 10 in Appendix B and use the entire sample of 100 plain M&M candies to construct a 95% confidence interval for the mean weight of all M&Ms. (*Hint*: It is not necessary to manually enter the 100 weights, because they are already stored in separate STATDISK files of blue.sdd, brown.sdd, and so on. Go to the Sample Editor module and click on the Help bar. Carefully read the instructions for combining data sets into one big data set. Follow those directions and combine the different M&M data sets into one big data set. That is, combine the files BLUE, BROWN, GREEN, ORANGE, RED, and YELLOW into one data set named M&M.) Now use the sample statistics to construct a 95% confidence interval estimate of the population mean of all IQ scores. Enter the 95% confidence interval here.

_____

6-12. **Confidence Interval for Proportion** Exercise 22 in Section 6-5 of *Elementary Statistics* states that in a recent presidential election, 611 voters were surveyed and 308 of them said that they voted for the candidate who won (based on data from the ICR Survey Research Group). Find a 98% confidence interval estimate of the *percentage* of voters who said that they voted for the candidate who won. Of those who voted, 43% actually voted for the candidate who won. Is this result consistent with the survey results? How might a discrepancy be explained?

_____

6-13. **Limitations on Proportions** Assume that 150 college students are asked if they have ever purchased a textbook, and 150 of them answer "yes." With $n = 150$ and $x = 150$, try using STATDISK to construct a 95% confidence interval for the population proportion of all students who answer "yes." What happens?

_____

Try manually constructing the 95% confidence interval by using the procedures described in Section 6-5 of *Elementary Statistics*. What happens?

_____

Examine the assumptions listed at the beginning of Section 6-5. Are any of them violated with the sample data given in this experiment? If so, identify the violated assumptions.

_____

_____

6-14. **Finding the Number of Successes** The *Columbia Daily Monitor* reports that when 1255 people were surveyed, 24% said that they could identify the name of the county superintendent. Given that 24% of the 1255 people surveyed could correctly identify the name in question, what is the actual number of people who could make that correct identification? How did you find this number?

_____

_____

Use STATDISK to construct a 95% confidence interval estimate of the percentage of people who could correctly identify the name of the county superintendent.

_____

Write a statement that correctly *interprets* the above confidence interval.

_____

6-15. **Finding the Number of Successes** Exercise 24 in Section 6-5 of *Elementary Statistics* states that in a study of the Clark method of gender selection, 40 couples tried to have baby girls. Among the 40 babies, 62.5% were girls. First find the actual number of girls in the sample, then use STATDISK to construct a 95% confidence interval estimate for the proportion of girls from all couples who try to have baby girls with the Clark method of gender selection. Based on the result, can we conclude that the Clark method is effective with a rate of girls that is greater than 50%?

_____

_____

6-16. **Finding the Number of Successes** Exercise 26 in Section 6-5 of *Elementary Statistics* states that the tobacco industry closely monitors all surveys that involve smoking. One survey showed that among 785 randomly selected subjects who completed four years of college, 18.3% smoke (based on data from the American Medical Association). Find the number of survey subjects who smoke, then use STATDISK to construct the 98% confidence interval for the true percentage of smokers among all people who completed four years of college.

_____

Based on the above result, does the smoking rate for those with four years of college appear to be substantially different than the 27% rate for the general population? Explain.

_____

Write a statement that correctly *interprets* the confidence interval obtained above.

_____

6-17. **Sample Size for Estimating Mean** Exercise 2 in Section 6-4 of *Elementary Statistics* states that the Franklin Vending Machine Company must adjust its machines to accept only coins with specified weights. We want to obtain a sample of quarters and weigh them to determine the mean. Use STATDISK to determine how many quarters must be randomly selected and weighed if we want to be 99% confident that the sample mean is within 0.025 g of the true population mean for all quarters. If we use the sample of quarters in Data Set 13 of Appendix B of the textbook, we can estimate that the population standard deviation is 0.068 g.

      Sample size: _____

6-18. **Sample Size for Estimating Mean** Exercise 4 in Section 6-4 of *Elementary Statistics* states that we want to estimate the mean weight of plastic discarded by households in one week. Use STATDISK to determine how many households must be randomly selected if we want to be 99% confident that the sample mean is within 0.250 lb of the true population mean. Data Set 5 in Appendix B includes the weights of plastic discarded for 62 households (based on data from the Garbage Study at the University of Arizona). If we use that sample as a pilot study, we obtain a standard deviation of $s =$ 1.065 lb.

      Sample size: _____

6-19. a. Use STATDISK to find the sample size required to estimate the mean IQ of all college students. Assume that we want 95% confidence that the sample mean is in error by no more than 1.5 IQ points. (The value of the standard deviation σ for IQ scores of college students is not known, however we do know that $\sigma = 15$ for the general population. College students are a more homogeneous group, so σ should be less than 15, but if we let $\sigma = 15$, we are playing it safe by using a sample size larger than necessary.) Enter the sample size below.

      Sample size: _____

continued

b.   What happens to the sample size as the level of confidence is increased above 95%? Why?

_____

_____

_____

c.   If the degree of confidence is held constant, but the margin of error is increased from the current value of 1.5 IQ points, how is the sample size affected? Why?

_____

_____

_____

6-20. **Sample Size for Proportion: Using Prior Information** Exercise 27 from *Elementary Statistics* states that a researcher wants to estimate the percentage of students aged 12-18 who use computers in school. How many randomly selected students must be surveyed if she wants to be 98% confident that the margin of error is five percentage points? Use STATDISK to find the sample size with each of the following assumptions.

Assume that we have an estimate of $\hat{p}$ found from a prior study which revealed a percentage of 82% (based on data from the Consumer Electronics Manufacturers Association).

_____

Assume that we have no prior information suggesting a possible value of $\hat{p}$.

_____

Does the prior information have much of an effect on the sample size?

_____

6-21. **Sample Size for Proportion: Using Prior Information** Exercise 28 from *Elementary Statistics* states that the Spalding Corporation wants to estimate the proportion of golfers who are left-handed. (The company can use this information in planning for the number of right-handed and left-handed sets of golf clubs to make.) How many golfers must be surveyed if we want 99% confidence that the sample proportion has a margin of error of 0.025? Use STATDISK to find the sample size using each of the following two assumptions.

Assume that there is no available information that could be used as an estimate of $\hat{p}$.

_____

Assume that we have an estimate of $\hat{p}$ found from a previous study which suggests that 15% of golfers are left-handed (based on a *USA Today* report).

_____

6-22. **Effect of Sample Size** In this experiment we will study the effect of sample size on the confidence interval.

a. Begin by constructing the 95% confidence interval for the true proportion of successes given the following sample results: The number of trials is $n = 100$ and the number of successes is $x = 30$. Use STATDISK to find this confidence interval and enter the result here. Note that the sample proportion given here is $30/100 = 0.3$.

_____

b. Construct the 95% confidence interval for the true proportion of successes given the following sample results: The number of trials is $n = 200$ and the number of successes is $x = 60$. Use STATDISK to find this confidence interval and enter the result here. Note that the sample proportion given here is $60/200 = 0.3$ so this sample proportion is the same as it was in part (a).

_____

c. Construct the 95% confidence interval for the true proportion of successes given the following sample results: The number of trials is $n = 400$ and the number of successes is $x = 120$. Use STATDISK to find this confidence interval and enter the result here. Note that the sample proportion given here is $120/400 = 0.3$ so this sample proportion is the same as it was in parts (a) and (b).

_____

d. Construct the 95% confidence interval for the true proportion of successes given the following sample results: The number of trials is $n = 800$ and the number of successes is $x = 240$. As in the preceding parts of this experiment, the sample proportion is again equal to 0.3. Enter the confidence interval here.

_____

e. Construct the 95% confidence interval for the true proportion of successes given the following sample results: The number of trials is $n = 1600$ and the number of successes is $x = 480$. As in the preceding parts of this experiment, the sample proportion is again equal to 0.3. Enter the confidence interval here.

_____

f. Parts (a) through (e) of this experiment had a sample proportion of 0.3 and the degree of confidence was kept constant at 95%. Identify the effect of increasing the sample size.

_____

g. Explain why the confidence interval limits change, even though the sample proportion of 0.3 is kept constant.

_____

6-23. **Estimating Population Standard Deviation** Refer to the sample data used in Experiment 6-1 and use STATDISK to construct a 95% confidence interval to estimate the population standard deviation σ. Enter the result here.

_____

Refer to Section 6-6 in *Elementary Statistics* and identify the assumptions for the procedures used to construct confidence intervals for estimating a population standard deviation.

List those assumptions here.

_____

_____

Assuming that the sample is a simple random sample, how can the other assumption be checked? Can STATDISK be used to check for normality? If so, do such a check and report the results and conclusion here.

_____

_____

6-24. **Sample Size for Standard Deviation** Exercises 9-12 in Section 6-6 of *Elementary Statistics* are given below. Use STATDISK to find the indicated sample sizes. In each case assume that the sample is a simple random sample obtained from a normally distributed population.

a. (Exercise 9) Find the minimum sample size needed to be 95% confident that the sample standard deviation s is within 5% of σ .

Sample size: _____

b. (Exercise 10) Find the minimum sample size needed to be 95% confident that the sample standard deviation s is within 20% of σ .

Sample size: _____

c. (Exercise 11) Find the minimum sample size needed to be 99% confident that the sample variance is within 10% of the population variance.

Sample size: _____

d. (Exercise 12) Find the minimum sample size needed to be 95% confident that the sample variance is within 10% of the population variance.

Sample size: _____

7

Hypothesis Testing

STATDISK is designed for the hypothesis tests included in Chapter 7 of *Elementary Staitistics*, as well as other chapters. Begin by clicking on he main menu item of **Analysis**, then select the subdirectory item of **Hypothesis Testing**. You will now see the following menu of choices.

Mean - One Sample

Mean - One Sample (Raw Data)

Mean - Two Independent Samples

Mean - Matched Pairs

Proportion - One Sample

Proportion - Two Samples

St. Dev. - One Sample

St. Dev. - Two Samples

Among the items in the above list, four include a reference to *one sample*, and they involve hypothesis tests with claims made about a single population, as discussed in Chapter 7 of the textbook. The other four items involve *two* sets of sample data as described in Chapter 8 of the textbook. This chapter will consider hypothesis testing involving only one sample. The claim may be made about the mean of a single population, or the proportion of a single population, or the standard deviation or variance of a single population. In Chapter 8 we consider claims involving two samples.

We use hypothesis testing when we want to test some claim made about a particular characteristic of some population. In addition to the claim itself, we also need sample data and a significance level.

7-1 Testing Hypotheses About μ

Sections 7-2, 7-3, and 7-4 of *Elementary Statistics* are devoted to procedures for testing claims about the mean of a single population. The textbook explains that there are different procedures, depending on the size of the sample, the nature of the population distribution, and whether the population standard deviation σ is known. The textbook includes Figure 7-11 in Section 7-4 to help in selecting the correct hypothesis-testing procedure. However, STATDISK greatly simplifies the whole process because it is programmed to make the correct choice, depending on the information that is supplied. (One exception: Like other statistics software packages, STATDISK's hypothesis testing modules are not programmed to check for normality of the population, so you should not use a t-test when you know that the population has a distribution that is very nonnormal.)

Section 7-3 of the textbook illustrates the test of the claim that $\mu = 98.6$ for the body temperatures of healthy adults. That test uses a 0.05 level of significance and the sample data are listed on the top of the following page. We will illustrate STATDISK's hypothesis-testing feature by using those sample temperatures to test the null hypothesis of H_0: $\mu = 98.6$.

```
98.6  98.6  98.0  98.0  99.0  98.4  98.4  98.4  98.4  98.6
98.6  98.8  98.6  97.0  97.0  98.8  97.6  97.7  98.8  98.0
98.0  98.3  98.5  97.3  98.7  97.4  98.9  98.6  99.5  97.5
97.3  97.6  98.2  99.6  98.7  99.4  98.2  98.0  98.6  98.6
97.2  98.4  98.6  98.2  98.0  97.8  98.0  98.4  98.6  98.6
97.8  99.0  96.5  97.6  98.0  96.9  97.6  97.1  97.9  98.4
97.3  98.0  97.5  97.6  98.2  98.5  98.8  98.7  97.8  98.0
97.1  97.4  99.4  98.4  98.6  98.4  98.5  98.6  98.3  98.7
98.8  99.1  98.6  97.9  98.8  98.0  98.7  98.5  98.9  98.4
98.6  97.1  97.9  98.8  98.7  97.6  98.2  99.2  97.8  98.0
98.4  97.8  98.4  97.4  98.0  97.0
```

In Section 1-2 of this manual/workbook we presented the method for entering data into STATDISK, and in Section 2-1 we described the procedure for obtaining descriptive statistics. If we enter the body temperature data listed above, we will obtain these descriptive statistics:

$$n = 106 \qquad \bar{x} = 98.20 \qquad s = 0.62$$

STATDISK allows us to use either these summary statistics or the original list of raw data.

STATDISK Procedure for Hypothesis Tests about a Mean

1. Select the main menu item of **Analysis**.

2. Select **Hypothesis Testing** from the subdirectory.

3. If the summary statistics of n, $\bar{x}$, and s are all known, select the first item of **Mean - One Sample**. If those statistics are not known and you prefer to use the original list of raw data, select the second item of **Mean - One Sample (Raw Data)**.

4. You will now have a dialog box for entry of the claim, significance level, value of the claimed mean, and entry of either the sample statistics or the raw data. Make these entries.

   -Select the format of the claim that is being tested. The default will appear as

      1) Pop. Mean = Claimed Mean

   and that can be changed to any of the other possibilities by using the mouse to scroll through the 6 options. Click on the box to make the other options appear, then click on the format of the claim being tested.
   -Enter a significance level, such as 0.05 or 0.01.
   -Enter the *claimed* value of the population mean.
   -Enter the *population* standard deviation σ if it is known. If σ is not known (as is usually the case), ignore that box and leave it empty. (*Caution*: Be careful to avoid the mistake of incorrectly entering the *sample* standard deviation in the box for the *population* standard deviation.)
   -Depending on whether you chose to enter the summary statistics or raw data, you should now enter the sample size n, the sample mean $\bar{x}$, and the sample standard deviation s, or you should manually enter the raw data (or you can use Copy/Paste to copy the data to this module).

-Click the **Evaluate** bar to get the test results.

-Click the **Plot** bar to get a graph that shows the test statistic and critical values.

Shown below is the STATDISK display that results from testing the null hypothesis H_0: $\mu = 98.6$ with the sample statistics of $n = 106$, $\bar{x} = 98.20$, and $s = 0.62$. The important elements of the display are the P-value of 0.0000, the test statistic of $z = -6.6423$, and the critical values of $z = \pm 1.96$. The P-value of 0.0000 indicates that the P-value is so small that when rounded to four decimal places, all of the digits are 0. This implies that the P-value is less than 0.00005. Note that the display also includes the 95% confidence interval, and the conclusion to reject the null hypothesis, along with the conclusion that the "sample provides evidence to reject the claim," which was $\mu = 98.6$.

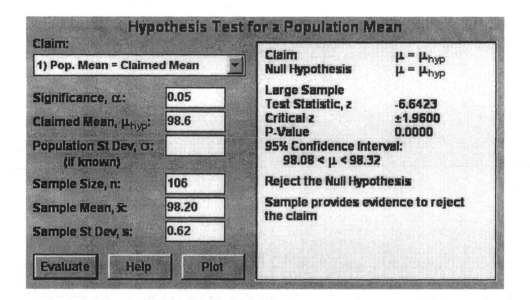

Because the above results include both the critical and P-values, you can use either the traditional approach of testing hypotheses (by comparing the test statistic to the critical values) or the P-value approach (by comparing the P-value to the significance level α).

If you click on the **Plot** bar, you will also obtain the graph shown on the top of the next page. Because the graph will appear in color, it will be easier to interpret on the computer monitor than the black and white reproduction shown on the following page. The vertical line representing the test statistic will be in blue, and the vertical lines representing the critical values will be in red.

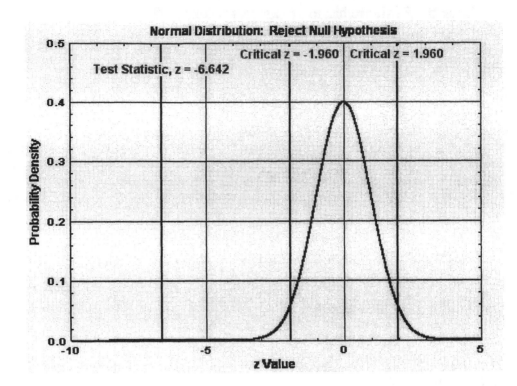

7-2 Testing Hypotheses About p

The STATDISK procedure for testing claims about a population proportion is quite easy, and it follows the same basic pattern described in Section 7-1 of this manual/workbook. In addition to having a claim to be tested, STATDISK also requires a significance level, the sample size n, and the number of successes x. In Section 6-2 of this manual/workbook, we briefly discussed one particular difficulty that arises when the available information provides the sample size n and the sample proportion $\hat{p}$ instead of the number of successes x. We provided this procedure for determining the number of successes x.

To find the number of successes x from the sample proportion and sample size:

Calculate $x = \hat{p} n$ and round the result to the nearest whole number.

We gave an example involving a 5.2% pilot death rate for 8411 crashes. We found that $x = (0.052)(8411) = 437.372$. Because x must be a whole number, we get $x = 437$.

Once the claim has been identified and the sample values of n and x have been determined, we can proceed to use STATDISK.

STATDISK Procedure for Testing Claims about *p*

1. Select **Analysis** from the main menu.

2. Select **Hypothesis Testing** from the subdirectory.

3. Select **Proportion - One Sample**.

4. Make these entries in the dialog box.

   -Select the format of the claim that is being tested. The default will appear as

   1) Pop. Proportion = Claimed Proportion

   and that can be changed to any of the other possibilities by using the mouse to scroll through the 6 options. Click on the box to make the other options appear, then click on the format of the claim being tested.

   -Enter a significance level, such as 0.05 or 0.01.

   -Enter the *claimed* value of the population proportion.

Section 7-5 of the textbook includes this example:

> ***Survey of Voters***   *In a survey of 1002 people, 701 said that they voted in the recent presidential election (based on data from ICR Research Group). Test the claim that when surveyed, the proportion of people who say that they voted is equal to 0.61, which is the proportion of people who actually did vote. Use a 0.05 significance level with the traditional method of testing hypotheses.*

From this statement we see that $n = 1002$, $x = 701$, the significance level is 0.05, and the claim is that $p = 0.61$. Using STATDISK with these values, we obtain the display shown on the top of the next page.

Important elements of the STATDISK display include the *P*-value of 0.0000, the test statistic of $z = 5.8150$, and critical values of ± 1.96. Having the *P*-value and critical values available, we can use either the traditional method of testing hypotheses or the *P*-value method. For this example, we know that the *P*-value is less than the significance level, so we reject the null hypothesis of $p = 0.61$. If we were to use the traditional approach, we see that the test statistic of $z = 5.8150$ exceeds the right critical value of $z = 1.96$, so we reject the null hypothesis of $p = 0.61$. There is sufficient evidence to warrant rejection of the claim that when surveyed, the proportion of people who say that they voted is equal to 0.61, which is the proportion of people who actually did vote.

We could again click on the **Plot** bar to obtain a graph showing the location of the test statistic and *P*-values.

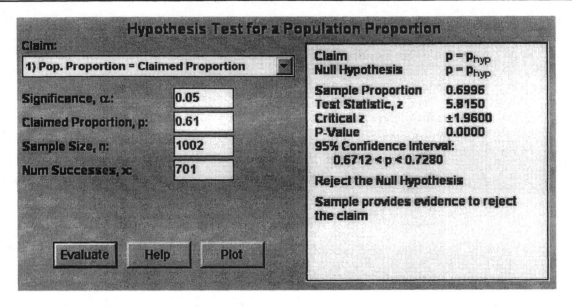

7-3 Testing Hypotheses About σ or σ^2

Caution: See Section 7-6 of *Elementary Statistics* where these two important assumptions are noted:

1.   The samples are simple random samples. (Remember the importance of good sampling methods.)

2.   The sample values come from a population with a *normal distribution*.

The textbook makes the very important point that tests of claims about standard deviations or variances are much stricter about the requirement of a normal distribution. If the population does not have a normal distribution, then inferences about standard deviations or variances can be very misleading. Suggestion: Given sample data, construct a histogram and/or normal probability plot to determine whether the assumption of a normal distribution is reasonable. If the population distribution does not appear to have a normal distribution, do not use the methods described in Section 7-6 of the textbook or this section of this manual/workbook. If the population distribution does appear to be normal and you want to test a claim about the population standard deviation or variance, use the STATDISK procedure given below.

STATDISK appears to work only with standard deviations because it requires entry of the sample standard deviation s and the claim must be stated in terms of the population standard deviation σ. However, claims about the population variance can be handled as well. For example, if you want to test the claim that $\sigma^2 = 9$, take the square root to restate the claim as $\sigma = 3$. Also, STATDISK requires entry of the sample *standard deviation s*, so if the sample variance is known, be sure to enter the square root.

STATDISK Procedure for Testing Hypotheses about σ or σ^2

1. Select **Analysis** from the main menu.

2. Select **Hypothesis Testing** from the subdirectory.

3. Select the option of **St Dev - One Sample**. (Select this option for claims about standard deviations or variances.)

4. Make these entries in the dialog box.

> -In the "claim box," select the format of the claim being tested.
>
> -Enter a significance level, such as 0.05 or 0.01.
>
> -Enter the *claimed* value of the standard deviation. (This is the value used in the statement of the null hypothesis.)
>
> -Enter the sample size n.
>
> -Enter the value of the sample standard deviation s.

This example is included in Section 7-6 of the textbook:

> ***IQ Scores of Statistics Professors*** *For a simple random sample of adults, IQ scores are normally distributed with a mean of 100 and a standard deviation of 15. A simple random sample of 13 statistics professors yields a standard deviation of s = 7.2. A psychologist is quite sure that statistics professors have IQ scores that have a mean greater than 100. He doesn't understand the concept of standard deviation very well and does not realize that the standard deviation should be lower than 15 (because statistics professors have less variation than the general population). Instead, he claims that statistics professors have IQ scores with a standard deviation equal to 15, the same standard deviation for the general population. Use a 0.05 level of significance to test the claim that $\sigma = 15$. Based on the result, what do you conclude about the standard deviation of IQ scores for statistics professors?*

From the above statements, we see that we want to test the claim that $\sigma = 15$, and we want to use a 0.05 significance level. Sample data consist of $n = 13$ and $s = 7.2$. The STATDISK displays are shown on the following page. When this example was discussed in Section 7-6 of the textbook, it was noted that the P-value is less than 0.01 (based on Table A-4), but the STATDISK display shows that the P-value is 0.0060. With STATDISK, it is much easier to find the P-value, so the P-value approach to hypothesis testing can be used with much greater ease. Using the P-value approach, we see that the P-value of 0.0060 is less than the significance level of 0.05, so we reject the null hypothesis that $\sigma = 15$. It appears that there is sufficient evidence to warrant rejection of the claim that statistics professors have IQ scores with a standard deviation that is equal to 15.

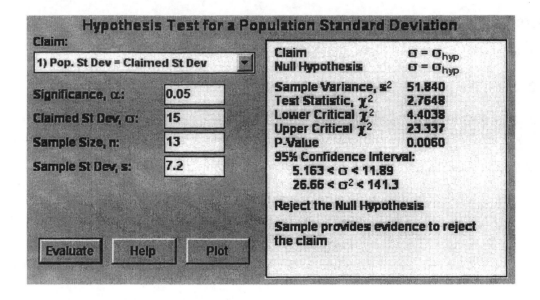

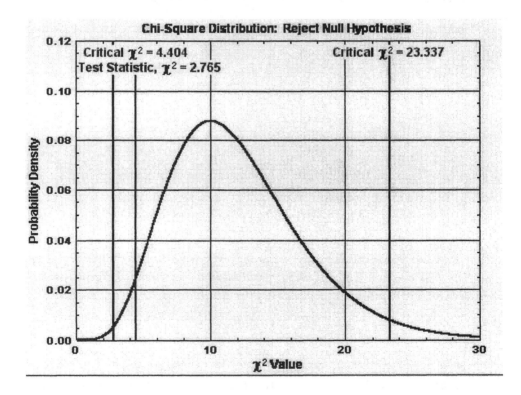

7-4   Hypothesis Testing with Simulations

Sections 7-1, 7-2, and 7-3 of this manual/workbook have all focused on using STATDISK for hypothesis tests using the traditional approach or *P*-value approach. Another very different approach is to use *simulations*. Let's illustrate the simulation technique with an example.

   In Section 7-1 of this manual/workbook and in Section 7-3 of the textbook, we tested the null hypothesis that $\mu = 98.6$ for the body temperatures of healthy adults. We used sample data consisting of 106 values having a mean $\bar{x} = 98.20$ and a standard deviation $s = 0.62$.

The key question is this:

> If the population mean body temperature is really 98.6, then how
> likely is it that we would get a sample mean of 98.20, given that
> the population has a normal distribution and the sample size is 106?

If the probability of getting a sample mean such as 98.20 is very small, then that suggests that the sample results are not explained as the result of chance random fluctuation. If the probability is high, then we can accept random chance as an explanation for the discrepancy between the sample mean of 98.20 and the claimed mean of 98.6. What we need is some way of determining the likelihood of getting a sample mean such as 98.20. That is the precise role of *P*-values in the *P*-value approach to hypothesis testing. However, there is another approach. STATDISK and many other software packages are capable of generating random results from a variety of different populations. Here is how STATDISK can be used: Determine the likelihood of getting a sample mean of 98.20 by randomly generating different samples from a population that is normally distributed with the claimed mean of 98.6. For the standard deviation, we will use the best available information: the value of $s = 0.62$ obtained from the sample.

STATDISK Procedure for Testing Hypotheses with Simulations

1. Identify the values of the sample size n, the sample standard deviation s, and the claimed value of the population mean.

2. Click on **Data**.

3. Click on **Normal Generator.**

4. Generate a sample randomly selected from a population with the claimed mean. When making the required entries in the dialog box, use the *claimed* mean, the sample size n, and the sample standard deviation s.

5. Continue to generate similar samples until it becomes clear that the given sample mean is or is not likely. (Here is one criterion: The given sample mean is *unlikely* if values at least as extreme occur 5% of the time or less.) If it is unlikely, reject the claimed mean. If it is likely, fail to reject the claimed mean.

For example, here are 10 results obtained from the random generation of samples of size 106 from a normally distributed population with a mean of 98.6 and a standard deviation of 0.62:

98.62	98.71	98.56	98.58	98.57
98.44	98.79	98.49	98.70	98.71

Examining the 10 sample means, we can see that they stay fairly close to 98.6. The largest deviation away from 98.6 is 0.19, compared to the deviation of 0.40 for the sample mean of 98.20. We can see from these sample values that a sample mean such as 98.20 is not at all typical for these circumstances. This suggests that 98.20 is a result that is *significantly* different. We would feel more confident in this conclusion if we had more sample results, so we could continue to randomly generate simulated samples until we feel quite confident in our thinking that a sample mean such as 98.20 is quite unusual. It is not likely to occur as the result of chance random variation. A better explanation is that the true population mean is not 98.6, as was claimed.

CHAPTER 7 EXPERIMENTS: Hypothesis Testing

7-1.   **Coke Volumes** Section 7-3 in *Elementary Statistics* includes an example of testing the claim that the mean volume of Coke in cans is greater than 12 oz. Use STATDISK and the given sample data ($n = 36$, $\overline{x} = 12.19$ oz, $s = 0.11$ oz) to test that claim. Use a significance level of 0.01. Enter the results here.

Test statistic:_____   Critical value(s):_____   *P*-value:_____

STATDISK conclusion about the null hypothesis:_____

STATDISK conclusion about the original claim:_____

Conclusion in your own words:_____

_____

7-2.   **BMW Crash Tests** Exercise 11 in Section 7-4 of *Elementary Statistics* involves crash tests involving a sample of five BMW cars. The repair costs (in dollars) are found to be as listed below. Use a 0.05 significance level to test the claim that the mean for all BMW cars is less than $1000.

797   571   904   1147   418

Test statistic:_____   Critical value(s):_____   *P*-value:_____

STATDISK conclusion about the null hypothesis:_____

STATDISK conclusion about the original claim:_____

Conclusion in your own words:_____

_____

Would BMW be justified in advertising that under the standard conditions, the repair costs are less than $1000?

_____

7-3.   **Retrieving Data: Second-Hand Smoke** Exercise 22 in Section 7-3 of *Elementary Statistics* refers to Data Set 9 in Appendix B, which includes the measured levels of serum cotinine (in ng/ml) in a sample of people who are nonsmokers and have no environmental tobacco smoke exposure at home or work (NOETS). Use STATDISK to test the claim of tobacco company spokesmen that these people have "positive" levels of cotinine, suggesting that this measure is not a good indicator of exposure to tobacco smoke. Use a significance level of 0.005. Instead of locating and manually entering the sample data, use File/Open to open the STATDISK file noets.sdd, which contains the sample data. Enter the results on the top of the next page.

Test statistic:_____ Critical value(s):_____ *P*-value:_____

STATDISK conclusion about the null hypothesis:_____

STATDISK conclusion about the original claim:_____

Conclusion in your own words:_____

7-4. **Retrieving Data: Textbook Prices** Exercise 24 in Section 7-3 of *Elementary Statistics* refers to Data Set 2 in Appendix B. Use STATDISK to test the claim that new textbooks at the University of Massachusetts have a mean less than $70. Use a 0.05 significance level. Instead of manually entering the sample data, use File/Open to open the STATDISK file newumass.sdd. Sample statistics: $n =$ _____; $\bar{x} =$ _____; $s =$ _____.

Test statistic:_____ Critical value(s):_____ *P*-value:_____

STATDISK conclusion about the null hypothesis:_____

STATDISK conclusion about the original claim:_____

Conclusion in your own words:_____

7-5. **Effect of the Significance Level** Repeat Experiment 7-4 using significance levels of 0.10, 0.01, and 0.02. Enter the results in the table below.

	0.10	0.05	0.02	0.01
Test Statistic				
Critical value(s)				
P-value				
Reject H_0?				

Based on the above table entries, how are hypothesis test results affected by decreasing the value of the significance level?

7-6. **Critical Mean** For Experiment 7-4, use trial and error with different values of the sample mean and identify the value of the sample mean that results in a *P*-value of 0.05. (Use the same sample standard deviation and sample size.) This is the value of the sample mean that corresponds to the critical value. Enter that value here:_____

As a sample mean moves farther away from the claimed value of the mean, what happens to the *P*-value?

As a sample mean moves farther away from the claimed value of the mean, what happens to the test statistic?

7-7. **Effect of Decreasing the Significance Level**   Refer to the claim and data in Experiment 7-4. Use trial and error with different values of the sample mean and identify the largest value of the sample mean that leads to rejection of the null hypothesis for the given significance levels. (Use the same sample standard deviation and sample size.) Enter the results below.

Significance Level	Largest Value of $\bar{x}$ that Leads to Rejection of the Null Hypothesis
0.10	
0.05	
0.02	
0.01	
0.005	
0.00001	

In general, as the significance level is made smaller, what can you conclude about the sample data required to reject the null hypothesis?

7-8. **Effect of Sample Size**   For experiment 7-4, use the same values of the sample mean and sample standard deviation, but change the sample size from 40 to 60, then 80, then 100. Enter the results in the table below.

	$n = 40$	$n = 60$	$n = 80$	$n = 100$
Test Statistic				
Critical value(s)				
P-value				
Reject H_0?				

As the sample size increases while the sample mean and sample standard deviation remain the same, what happens to the test statistic?

As the sample size increases while the sample mean and sample standard deviation remain the same, what happens to the P-value? Why does this happen?

7-9. **Effect of an Outlier**   Repeat Experiment 7-4 after changing the first sample value from $84.30 to $8430. The omission of a decimal point is a somewhat common error when entering data.

Test statistic:_____   Critical value(s):_____   P-value:_____

Conclusion in your own words:_____

In general, how are hypothesis test results affected by the presence of an outlier?

7-10. **Effect of Units of Measurement** Experiment 7-4 is based on values given in dollars. What happens if all of the sample values are changed to units of cents instead of dollars, while the claim is changed from "less than $70" to "less than 7000 cents"?

Test statistic:_____ Critical value(s):_____ P-value:_____

STATDISK conclusion about the null hypothesis:_____

STATDISK conclusion about the original claim:_____

Conclusion in your own words:_____

_____

In general, how are hypothesis test results affected by the units of measurement?

_____

7-11. **Effect of Standard Deviation** For the data used in Experiment 7-4, the sample standard deviation of the 40 textbook prices is $23.076. Use the same sample mean and sample size, but use these larger values of the sample standard deviation: $46 and $69. Enter the results below.

	_s_ = $23.076	_s_ = $46	_s_ = $69
Test Statistic			
Critical value(s)			
P-value			
Reject H_0?			

In general, how are hypothesis test results affected by increasingly large values of the sample standard deviation?

_____

7-12. **Telephone Use** Exercise 2 in Section 7-5 of *Elementary* Statistics refers to a survey of 4276 randomly selected households. Among those households, 4019 had telephones (based on data from the U.S. Census Bureau). Use STATDISK with those survey results to test the claim that the percentage of households is now greater than the 35% rate that was found in 1920. Use a 0.01 significance level.

Test statistic:_____ Critical value(s):_____ P-value:_____

STATDISK conclusion about the null hypothesis:_____

STATDISK conclusion about the original claim:_____

Conclusion in your own words:_____

_____

The current rate of 4019/4276 (or 94%) appears to be significantly greater than the 1920 rate of 35%, but is there sufficient evidence to support that claim?_____

7-13. **Driving Practices** Exercise 3 in Section 7-5 of *Elementary Statistics* states that the Hartford Insurance Company obtains a simple random sample of 850 drivers and finds that 544 of them change tapes or CDs while driving (based on data from *Prevention* magazine). Use STATDISK with a 0.02 significance level to test the claim that most drivers change tapes or CDs while driving.

Test statistic:_____   Critical value(s):_____   *P*-value:_____

Conclusion in your own words:_____

_____

7-14. **Finding Number of Successes** Exercise 10 in Section 7-5 of *Elementary Statistics* involves the drug Ziac that is used to treat hypertension. In a clinical test, 3.2% of 221 Ziac users experienced dizziness (based on data from Lederle Laboratories). Further tests must be done for adverse reactions that occur in at least 5% of treated subjects. Using STATDISK and a 0.01 significance level, test the claim that fewer than 5% of all Ziac users experience dizziness.

Test statistic:_____   Critical value(s):_____   *P*-value:_____

STATDISK conclusion about the null hypothesis:_____

STATDISK conclusion about the original claim:_____

Conclusion in your own words:_____

_____

7-15. **Finding Number of Successes** Exercise 12 in Section 7-5 of *Elementary Statistics* refers to a survey showing that among 785 randomly selected subjects who completed four years of college, 18.3% smoke (based on data from the American Medical Association). Use STATDISK with a 0.01 significance level to test the claim that the rate of smoking among those with four years of college is less than the 27% rate for the general population.

Test statistic:_____   Critical value(s):_____   *P*-value:_____

STATDISK conclusion about the null hypothesis:_____

STATDISK conclusion about the original claim:_____

Conclusion in your own words:_____

_____

7-16. **Variation of Statistics Test Scores** Exercise 9 in Section 7-6 of *Elementary Statistics* states that in the author's past statistics classes, test scores have a standard deviation equal to 14.1. One of his current classes now has 27 test scores with a standard deviation of 9.3. Use STATDISK with a 0.01 significance level to test the claim that this current class has less *variation* than past classes. Enter the results on the next page.

Test statistic:_____   Critical value(s):_____   *P*-value:_____

STATDISK conclusion about the null hypothesis:_____

STATDISK conclusion about the original claim:_____

Conclusion in your own words:_____

_____

Does a lower standard deviation suggest that the current class is doing better?

_____

7-17. **Variation in Bank Customer Waiting Times** Exercise 10 in Section 7-6 of *Elementary Statistics* states that with individual lines at its various windows, the Jefferson Valley Bank found that the standard deviation for normally distributed waiting times on Friday afternoons was 6.2 min. The bank experimented with a single main waiting line and found that for a simple random sample of 25 customers, the waiting times have a standard deviation of 3.8 min. On the basis of previous studies, we can assume that the waiting times are normally distributed. Use STATDISK with a 0.05 significance level to test the claim that a single line causes lower variation among the waiting times.

Test statistic:_____   Critical value(s):_____   *P*-value:_____

STATDISK conclusion about the null hypothesis:_____

STATDISK conclusion about the original claim:_____

Conclusion in your own words:_____

_____

Why would customers prefer waiting times with less variation? Does the use of a single line result in a shorter wait?

_____

7-18. **Effect of Sample Size** We want to test the claim that "the population proportion of successes equals 0.5." We will use a 0.05 level of significance. First conduct the test assuming that there are 55 successes among 100 observations. Then repeat the test assuming that there are 550 successes among 1000 observations. Note that in both cases, the proportion of successes is 0.55. Enter the results here.

	$n = 100, x = 55$	$n = 1000, x = 550$
Conclusion	_____	_____
Test statistic	_____	_____
Critical values	_____	_____
P-value	_____	_____

Based on a comparison of the above results, what do you conclude?

_____

7-19. **Hypothesis Testing with Simulations**   In the example of Section 7-1 in this manual/workbook, we listed 106 sample values that were used to test the claim that $\mu = 98.6$. We noted that those sample values can be described with these summary statistics: $n = 106$, $\bar{x} = 98.20$, $s = 0.62$. Using those same sample values, again test the claim that $\mu = 98.6$, but use a 0.10 significance level.

Test statistic:_____   Critical value(s):_____   *P*-value:_____

STATDISK conclusion about the null hypothesis:_____

STATDISK conclusion about the original claim:_____

Conclusion in your own words:_____

_____

Instead of conducting a formal hypothesis test, we will now consider another way of determining whether the sample mean of 98.20 is significantly different than the claimed value of 98.6. With a significance level of $\alpha = 0.10$, we will use this criterion:

The sample mean of 98.2 is significantly different if there is a 10% chance (or less) that a sample mean will differ from 98.6 by an amount of 0.4 or more.

Use the Normal Generator module to randomly generate a sample of 106 values from a normally distributed population with a mean of 98.6 and a standard deviation of 0.62. Find the mean of the generated sample and record it in the space below. Then generate another sample, and another sample, and continue until you have enough sample means to determine how often a result such as $\bar{x} = 98.20$ (or any other value that differs from 98.6 by 0.4 or more) will occur.

_____

_____

_____

Enter the number of generated samples:_____

How often was the sample mean at least 0.4 away from 98.6? _____

What is the proportion of trials in which the sample mean was at least 0.4 away from 98.6? _____

Based on these results, what do you conclude about the claim that $\mu = 98.6$? Explain.

_____

_____

7-20. **Hypothesis Testing with Simulations** In Experiment 7-4 of this manual/workbook, we used 40 sample values to test the claim that $\mu < \$70$. The sample values can be described with these summary statistics: $n = 40$, $\bar{x} = \$65.117$, $s = \$23.076$. Using those same sample values, again test the claim that $\mu < \$70$, but use a 0.10 significance level.

Test statistic:_____ Critical value(s):_____ *P*-value:_____

STATDISK conclusion about the null hypothesis:_____

STATDISK conclusion about the original claim:_____

Conclusion in your own words:_____

_____

Instead of the formal hypothesis test, we will now use a simulation technique for determining whether the sample mean of $65.117 is significantly below the claimed value of $70. With a significance level of $\alpha = 0.10$, we will use this criterion:

The sample mean of $65.117 is significantly low if there is a 10% chance (or less) that a sample mean will be below the claimed value of $70 by an amount of $4.883 or more.

Use STATDISK's Normal Generator module to randomly generate a sample of 40 values from a normally distributed population with a mean of 70 and a standard deviation of 23.076. Find the mean of the generated sample and record it in the space below. Then generate another sample, and another sample, and continue until you have enough sample means to confidently determine how often a result such as $\bar{x} = \$65.117$ (or any other value lower than $65.117) will occur.

_____

_____

_____

Enter the number of generated samples:_____

How often was the sample mean equal to or less than $65.117?_____

What is the proportion of trials in which the sample mean was equal to or less than $65.117?_____

Based on these results, what do you conclude about the claim that $\mu < \$70$? Explain.

_____

_____

8

Inferences from Two Samples

8-1 Introduction

As the use of technology grows in introductory statistics courses, more instructors are expanding the scope of topics covered. They often accomplish this with assignments from chapters not formally covered in class. Once students understand the basic concepts of confidence interval construction and hypothesis testing, it becomes relatively easy to use technology like STATDISK to apply that understanding to other circumstances, such as those involving two samples instead of only one.

This chapter deals with inferences based on two sets of sample data. Some of the most important applications of statistics require the methods of this chapter, such as determining whether the proportion of adverse reactions in a sample of people using a new drug is the same as the proportion of adverse reactions in a sample of people using a placebo.

In this chapter we deal with the STATDISK modules relevant to Chapter 8 in *Elementary Statistics*. The sections of this chapter correspond to the sections in the textbook, so that Section 8-2 in this manual/workbook addresses the same topic as Section 8-2 in the textbook, Section 8-3 here matches Section 8-3 in the textbook, and so on. If you select **Analysis** from the main menu, then select the subdirectory item of **Hypothesis Testing**, you get the following options, which are listed along with the corresponding sections in the textbook.

STATDISK Hypothesis Tests	Sections in *Elementary Statistics*
Mean - One Sample	7-2, 7-3, 7-4
Mean - One Sample (Raw Data)	7-2, 7-3, 7-4
Mean - Two Independent Samples	8-2, 8-6
Mean - Matched Pairs	8-3
Proportion - One Sample	7-5
Proportion - Two Samples	8-4
St. Dev. - One Sample	7-6
St. Dev. - Two Samples	8-5

Confidence Intervals from Two Samples

If you examine only the STATDISK menu titles for Hypothesis Tests and Confidence Intervals, it might appear that there is no way to obtain confidence intervals based on two sets of sample data, but this is not the case. Here is an important point:

> **To obtain confidence intervals for cases with two samples, use the** *Hypothesis Testing* **option and test a claim of** *equality* **between the two parameters. A confidence interval will be included with the results.**

For example, to obtain a 95% confidence interval for the difference $\mu_1 - \mu_2$, use a 0.05 significance level with STATDISK's Hypothesis Testing feature to test the claim that $\mu_1 = \mu_2$. *The 95% confidence interval limits will be included with the displayed results.*

8-2 Two Means: Independent and Large Samples

Elementary Statistics has two sections in Chapter 8 devoted to inferences about means based on two independent samples. Section 8-2 considers only cases in which both samples are large (greater than 30), and Section 8-6 deals with cases involving small samples. There are some important issues to consider with small samples, but the case involving large samples is fairly straightforward. We describe the STATDISK procedure, then consider the example included in Section 8-2 of the textbook.

STATDISK Procedure for Inferences (Hypothesis Tests and Confidence Intervals) about Two Means: Independent and Large Samples

Use the following procedure for both hypothesis tests and confidence intervals. To obtain confidence intervals for cases with two large and independent samples, use the *Hypothesis Testing* option and test a claim of *equality* between the two parameters. A confidence interval will be included with the results.

1. For each of the two samples, identify the values of the sample sizes, sample means, and sample standard deviations. That is, find the values of n_1, $\bar{x}_1$, s_1, n_2, $\bar{x}_2$, and s_2. (If necessary, use STATDISK's Descriptive Statistics module to find the required sample statistics. See Section 2-1 of this manual/workbook.)

2. Select **Analysis** from the main menu.

3. Select **Hypothesis Testing** from the submenu.

4. Choose the option of **Mean - Two Independent Samples**.

5. Enter the required values in the dialog box. Be particularly careful with these items:

    -Claim: Be sure to select the form of the claim to be tested. If the objective is to find a confidence interval, choose the default option of Pop. mean 1 = Pop. Mean 2.

    -Avoid confusion between the *sample* standard deviation and the *population* standard deviation. Values of the population standard deviation are rarely know, so those boxes are usually left blank.

6. Click on the **Evaluate** button.

7. If a graph is desired, click on **Plot**.

Section 8-2 of the textbook includes an example involving a test of the claim that the mean weight of regular Coke is different from the mean weight of regular Pepsi. The summary statistics for the two samples are shown on the top of the next page. We want to use STATDISK with a 0.01 significance level to test the claim that the mean weight of regular Coke is different from the mean weight of regular Pepsi.

	Regular Coke	Regular Pepsi
n	36	36
$\bar{x}$	0.81682	0.82410
s	0.007507	0.005701

Hypothesis Test: After selecting **Analysis**, then **Hypothesis Testing**, and **Means - Two Independent Samples** as described in the above procedure, we proceed to make the required entries in the dialog box. Shown below is the dialog box with the entries corresponding to the two sets of sample data we are considering. After clicking on **Evaluate**, the displayed results appear as shown in the rightmost portion of the display.

Confidence Interval: Select **Analysis**, **Hypothesis Testing**, and **Means - Two Independent Samples**. In the dialog box, select the claim that has the form of "Pop. Mean 1 = Pop. Mean 2." Enter the significance level corresponding to the degree of confidence; enter a significance level of 0.01 for a 99% confidence interval. The confidence interval will be included among the displayed results, as shown below.

Hypothesis Test for the Mean of Two Independent Samples

Claim:

4) Pop. Mean 1 not = Pop. Mean 2

Claim	μ_1 not = μ_2
Null Hypothesis	$\mu_1 = \mu_2$

Significance, α: 0.01

Sample 1:

Sample Size, n_1: 36

Sample Mean, $\bar{x}_1$: 0.81682

Sample St Dev, s_1: 0.007507

Population St Dev, σ_1:
(if known)

Sample 2:

Sample Size, n_2: 36

Sample Mean, $\bar{x}_2$: 0.82410

Sample St Dev, s_2: 0.005701

Population St Dev, σ_2:
(if known)

Large Samples
Test Statistic, z   -4.6338
Critical z   ±2.5758
P-Value   0.0000
99% Confidence Interval:
   -0.01133 < μ_1-μ_2 < -0.003233

Reject the Null Hypothesis

Sample provides evidence to support the claim

Evaluate   Help   Plot

8-3 Two Means: Matched Pairs

Section 8-3 of *Elementary Statistics* describes methods for testing hypotheses and constructing confidence interval estimates of the differences between samples consisting of *matched pairs*. An example of data consisting of matched pairs can be seen in Table 8-1 from the textbook. Each pair of reported/measured heights is matched according to the person from whom the measurements were obtained. In the textbook we deleted the sample values for Student C on the basis that they constitute an outlier, which appears to be an error.

Table 8-1 Reported and Measured Heights of Male Statistics Students

Student	A	B	C	D	E	F	G	H	I	J	K	L
Reported Height	68	74	**82.25**	66.5	69	68	71	70	70	67	68	70
Measured Height	66.8	73.9	**74.3**	66.1	67.2	67.9	69.4	69.9	68.6	67.9	67.6	68.8
Difference	1.2	0.1	**7.95**	0.4	1.8	0.1	1.6	0.1	1.4	**-0.9**	0.4	1.2

STATDISK Procedure for Inferences (Hypothesis Tests and Confidence Intervals) about Two Means: Matched Pairs

To conduct hypothesis tests or construct confidence intervals based on data consisting of matched pairs, use the following procedure. (Again note that the construction of confidence intervals can be accomplished by using the **Hypothesis Testing** module; simply select a test of equality, and a confidence interval will be included among the displayed results.)

1.   Select **Analysis** from the main menu.

2.   Select **Hypothesis Testing** from the subdirectory.

3.   Select **Mean - Matched Pairs**.

4.   Make these entries and selections in the dialog box:

   -If testing a claim, select the appropriate format in the "claim" box. The default is Pop. Mean of Difference = 0, which is equivalent to $\mu_d = 0$. Scroll through the other 5 possibilities and select the format corresponding to the claim being tested. If you want to construct a confidence interval, proceed as if testing a claim and use the default of "Pop. Mean of Difference = 0."

   -Enter a significance level, such as 0.05 or 0.01.

   -Get the paired data into the two columns in the window. Either manually enter the data (if the list is not very long) or use Copy/Paste to copy the data from Sample Editor (after using Open to retrieve the data). If using Copy/Paste, copy one data set at a time.

   -Click the **Evaluate** bar.

   -Obtain a graph by clicking the **Plot** bar. The graph will show the test statistic and critical value(s).

Hypothesis Test: Refer to the sample data in Table 8-1 and *delete the values for Student C*. Following the above procedure, we make the entries in the dialog box as shown below. Note that the format of the claim corresponds to the claim that male students *exaggerate* their heights with reported values that are *greater than* the measured values. The results include the test statistic of $t = 2.7014$ and the *P*-value of 0.0111. Because the *P*-value is less than the significance level of 0.05, we reject the null hypothesis and support the claim that male students exaggerate their heights.

Confidence Interval: For the claim, select the default option of "Pop. Mean of Difference = 0" instead of the claim shown in the dialog box below. With a claim of "Pop. Mean of Difference = 0," the 95% confidence interval will be displayed as $0.1179 < \mu_d < 1.228$.

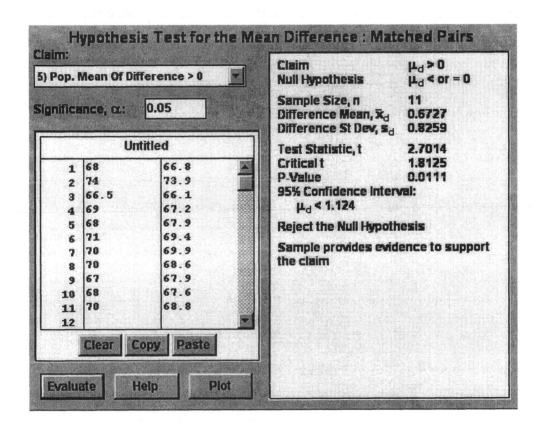

8-4 Two Proportions

The textbook makes the point that "a strong argument could be made that this section is one of the most important sections in the book because the main objective is to provide methods for dealing with two sample proportions - a situation that is very common in real applications."

STATDISK requires that you first identify the number of successes x_1 and the sample size n_1 for the first sample, and you must also know the values of x_2 and n_2 for the second sample. However, sample data often consist of sample proportions instead of the actual numbers of successes. Be sure to read the textbook carefully for a way to determine the number of successes. See the statement in Section 8-4 that "when 734 men were treated with

Viagra, 16% of them experienced headaches." From that statement we can see that $n_1 = 734$ and $\hat{p}_1 = 0.16$, but the actual number of successes x_1 is not given. However, from $\hat{p}_1 = x_1/n_1$, we know that $x_1 = n_1 \cdot \hat{p}_1$ so that $x_1 = 734 \cdot 0.16 = 117.44$. Because you cannot have 117.44 men who experienced headaches, x_1 must be a whole number and we round it to 117. We can now use $x_1 = 117$ in the calculations that require its value. In general, given a sample proportion and sample size, calculate the number of successes by multiplying the decimal form of the sample proportion and the sample size.

STATDISK Procedure for Inferences (Hypothesis Tests and Confidence Intervals) for Two Proportions

To conduct hypothesis tests or construct confidence intervals for two population proportions, use the following procedure. (Again note that the construction of confidence intervals can be accomplished by using the **Hypothesis Testing** module; simply select a test of equality, and a confidence interval will be included among the displayed results.)

1. For each of the two samples, find the sample size n and the number of successes x.

2. Select **Analysis** from the main menu.

3. Select **Hypothesis Testing** from the subdirectory.

4. Select **Proportion - Two Samples**.

5. Make these entries in the dialog box:

   -In the "claim" box, select the format corresponding to the claim. (If you want to construct a confidence interval, select the default option of "Pop. Proportion = Claimed Proportion.")

   -Enter a significance level, such as 0.05 or 0.01.

   -For each sample, enter the sample size n and the number of successes x.

6. Click on the **Evaluate** bar to obtain the test results.

7. Click on the **Plot** bar to obtain a graph that includes the test statistic and critical value(s).

Section 8-4 of the textbook includes this example:

> **Viagra Treatment and Placebo** The drug Viagra has become quite well-known, and it has had a substantial economic impact on its producer, Pfizer Pharmaceuticals. In preliminary tests for adverse reactions, it was found that when 734 men were treated with Viagra, 16% of them experienced headaches. (There's some real irony there.) Among 725 men in a placebo group, 4% experienced headaches. These results were provided by Pfizer Pharmaceuticals. Is there sufficient evidence to support the claim that among those men who take Viagra, headaches occur at a rate that is greater than the rate for those who do not take Viagra? Use a 0.01 level of significance.

The first step is to find the required numbers of successes, as described in the textbook. They are found as follows:

Viagra: 16% of 734 = $0.16 \times 734 = 117.44 = 117$ (rounded)

Placebo: 4% of 725 = 29

Hypothesis Test: We can now proceed to use the above STATDISK procedure. After selecting **Analysis, Hypothesis Testing, Proportion - Two Samples,** we make the required entries in the dialog box as shown below. Note the form of the claim used to test the claim that among those men who take Viagra, headaches occur at a rate that is *greater than* the rate for those who do not take Viagra.

Confidence Interval: For the claim, select the default option of "Pop. Proportion 1 = Pop. Proportion 2" instead of the claim shown in the dialog box below. With a claim of "Pop. Proportion 1 = Pop. Proportion 2," and with a significance level of 0.01, the 99% confidence interval will be displayed as $0.07987 < p_1 - p_2 < 0.1589$.

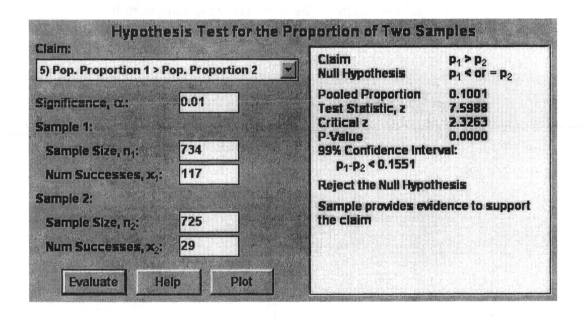

8-5 Two Variances

Section 8-5 in the textbook describes the use of the F distribution in testing a claim that two populations have the same variance (or standard deviation). STATDISK has a module for such tests. Section 8-5 in the textbook focuses on hypothesis tests, and confidence intervals are included only as an exercise (Exercise 15), but STATDISK will include confidence intervals for the ratio σ_1^2 / σ_2^2 and also for the ratio σ_1 / σ_2. Use the following procedure.

STATDISK Procedure for Inferences (Hypothesis Tests and Confidence Intervals) for Two Standard Deviations or Two Variances

  1.   For each of the two samples, obtain the sample size n, and the sample standard

deviation s. (If you have two samples of raw data, you can find these statistics using the Descriptive Statistics module, as described in Section 2-1 of this manual/workbook. Also, if the available information includes sample sizes and sample *variances*, be sure to take the square root of the sample variances to obtain the sample *standard deviations*.)

2.   Select **Analysis** from the main menu.

3.   Select **Hypothesis Testing** from the subdirectory.

4.   Select **St Dev - Two Samples.**

5.   Make these entries in the dialog box:

-In the "claim" box, select the case corresponding to the claim being tested.

-Enter a significance level, such as 0.05 or 0.01.

-In the appropriate boxes, enter the sample size and sample standard deviation for the first sample, then do the same for the second sample.

6.   Click on the **Evaluate** bar to obtain the test results.

7.   Click on the **Plot bar** to obtain a graph that includes the test statistic and critical value(s).

Section 8-5 in *Elementary Statistics* includes an example based on the following two sets of sample statistics.

	Regular Coke	Regular Pepsi
n	36	36
$\overline{x}$	0.81682	0.82410
s	0.007507	0.005701

The textbook describes the procedure for using a 0.05 significance level in testing the claim that the two populations of regular Coke and regular Pepsi have the same standard deviation. Using the above STATDISK procedure, the dialog box will be as shown on the next page. The textbook procedure requires that the sample with the larger variance be designated as Sample 1, but this is not necessary with STATDISK. Using the procedure in the textbook, the test statistic will always be at least 1; using STATDISK it possible to get test statistics less than 1. However, the final conclusions will always be the same. STATDISK automatically does the required calculations and it correctly handles cases in which the first sample has a variance smaller than the second sample.

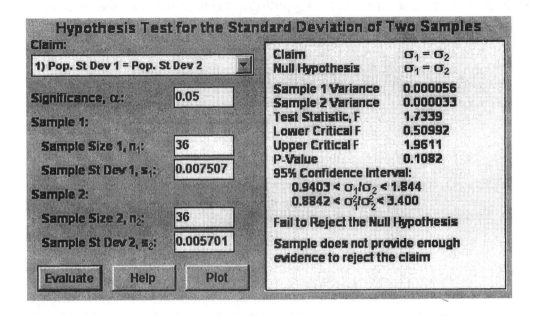

8-6   Two Means: Independent and Small Samples

The textbook considers inferences about means of independent samples by using two separate sections: Section 8-2 covers cases in which both samples are large ($n_1 > 30$) and ($n_2 > 30$); Section 8-6 discusses the following three cases in which the samples are not both large:

Case 1:	The values of both population variances are known. (In reality, this case rarely occurs.)
Case 2:	The two populations have equal variances. (That is, $\sigma_1^2 = \sigma_2^2$.)
Case 3:	The two populations have unequal variances. (That is, $\sigma_1^2 \neq \sigma_2^2$.)

If the two samples are both large or if both population standard deviations are known, STATDISK will automatically proceed and display the results. If either of the two samples are small and both populations standard deviations are not known, STATDISK will give you these options:

Eq. Vars: POOL	Assuming that $\sigma_1^2 = \sigma_2^2$, pool the two sample variances to obtain an estimate of the common population variance. (See Case 2 in the textbook.)
Not Eq. Vars: NO POOL	Assuming that $\sigma_1^2 \neq \sigma_2^2$, do not pool the sample variances. (See Case 3 in the textbook.)
Prelim. F test	Let STATDISK do an F test to decide whether $\sigma_1^2 = \sigma_2^2$. Based on the result, the procedures of Case 2 or Case 3 will be used. (Both cases are described in the textbook.)

Be sure to read the textbook discussion in Section 8-6 regarding the method to be used. Consult your instructor who might have a preferred method to be used.

STATDISK Procedure for Inferences (Hypothesis Tests and Confidence Intervals) about Two Means: Independent and Small Samples

1. For each of the two samples, find the sample size n, the sample mean $\bar{x}$, and the sample standard deviation s. (If you have a collection of raw data, the values of n, $\bar{x}$, and s can be found by entering the data in the sample editor, then using Copy/Paste to copy the data to the Descriptive Statistics module. See Section 2-1 of this manual/workbook for the detailed procedure.)

2. Select **Analysis** from the main menu.

3. Select **Hypothesis Testing** from the subdirectory.

4. Select **Mean - Two Independent Samples**.

5. Make these entries in the dialog box:

   -If testing a claim, select the appropriate form in the first entry box. That is, scroll through the 6 different possibilities and select the one corresponding to the given claim. If you simply want to construct a confidence interval for the difference $\mu_1 - \mu_2$, leave the claim box with the default of "Pop. Mean 1 = Pop. Mean 2.

   -Select a significance level, such as 0.05 or 0.01.

   -Proceed to enter the sample statistics for each of the two samples. In both cases, there is a box for entering the *population* standard deviation if it is known. If not known (as is usually the case), leave those boxes empty.

6. Click on the **Evaluate** bar.

7. If given a choice of a preliminary F test, pooling the sample variances, or not pooling the sample variances, make that choice. (See Section 8-6 in *Elementary Statistics*.)

8. Click on the **Plot** bar to get a graph that includes the test statistic and critical value(s).

Note: In Section 8-6 of *Elementary Statistics*, if you are dealing with two means and at least one of the samples is small and the two variances do not appear to be equal (Case 3), the textbook indicates that the number of degrees of freedom for the t test statistic is the smaller of $n_1 - 1$ and $n_2 - 1$. The textbook states that this is a more conservative and simplified alternative to computing the number of degrees of freedom by using Formula 8-1, which is given in Section 8-6 of the textbook. STATDISK uses Formula 8-1 instead of the smaller of $n_1 - 1$ and $n_2 - 1$. Consequently, the critical values and P-value generated by STATDISK will be different from those given in the textbook. The STATDISK results are better because they use Formula 8-1 for determining the number of degrees of freedom.

CHAPTER 8 EXPERIMENTS: Inferences from Two Samples

8-1. **Regular Coke/Diet Coke** Exercise 8-5 in Section 8-2 of *Elementary Statistics* includes the summary statistics for weights (in pounds) of regular Coke and diet Coke given below. Use STATDISK with a 0.01 significance level to test the claim that cans of regular Coke and diet Coke have the same mean weight.

$$\underline{\text{Regular Coke}} \qquad \underline{\text{Diet Coke}}$$
$$n_1 = 36 \qquad\qquad n_2 = 36$$
$$\bar{x}_1 = 0.81682 \qquad \bar{x}_2 = 0.78479$$
$$s_1 = 0.007507 \qquad s_2 = 0.004391$$

Test statistic:_____ Critical value(s):_____ *P*-value:_____

STATDISK conclusion about the null hypothesis:_____

STATDISK conclusion about the original claim:_____

Conclusion in your own words:_____

_____

Construct a 99% confidence interval estimate of $\mu_1 - \mu_2$, the difference between the mean weight of regular Coke and the mean weight of diet Coke.

_____

Write a statement that *interprets* that confidence interval.

_____

8-2. **M&Ms** Exercise 6 in Section 8-2 of *Elementary Statistics* states that Data Set 10 in Appendix B contains weights of 100 randomly selected M&M plain candies. Those weights have a mean of 0.9147 g and a standard deviation of 0.0369 g. A previous edition of this book used a different sample of 100 M&M plain candies (before blue was introduced), with a mean and standard deviation of 0.9160 g and 0.0433 g, respectively. Use STATDISK to determine whether there is sufficient evidence to conclude that the two population means are different.

Test statistic:_____ Critical value(s):_____ *P*-value:_____

STATDISK conclusion about the null hypothesis:_____

STATDISK conclusion about the original claim:_____

Conclusion in your own words:_____

_____

Construct a 99% confidence interval estimate of $\mu_1 - \mu_2$, the difference between the means of the two populations of weights.

_____

8-3. **Retrieving Data** Exercise 14 in Section 8-2 of *Elementary Statistics* refers to Data Set 1 in Appendix B. Instead of manually entering the data, use STATDISK to retrieve the file for the weights of diet Coke (ckdietwt.sdd) and the file for the weights of diet Pepsi (ppdietwt.sdd). Is there sufficient evidence to support a claim that the mean weight of cans of diet Coke is equal to the mean weight of cans of diet Pepsi?

Test statistic:_____   Critical value(s):_____   *P*-value:_____

STATDISK conclusion about the null hypothesis:_____

STATDISK conclusion about the original claim:_____

Conclusion in your own words:_____

_____

Construct a 99% confidence interval estimate of $\mu_1 - \mu_2$, the difference between the mean weight of diet Coke and the mean weight of diet Pepsi.

_____

Write a statement that *interprets* the confidence interval.

_____

8-4. **Retrieving Data** Exercise 15 in Section 8-2 of *Elementary Statistics* refers to Data Set 9 in Appendix B. Instead of manually entering the sample data, use STATDISK to retrieve the file for the serum cotinine levels of nonsmokers who are not exposed to smoke (noets.sdd) and the file for the serum cotinine levels of nonsmokers who are exposed to smoke (ets.sdd). Test the claim that both samples come from populations with the same mean.

Test statistic:_____   Critical value(s):_____   *P*-value:_____

STATDISK conclusion about the null hypothesis:_____

STATDISK conclusion about the original claim:_____

Conclusion in your own words:_____

_____

Construct a 99% confidence interval estimate of $\mu_1 - \mu_2$, the difference between the mean levels of cotinine in the two populations.

_____

Write a statement that *interprets* that confidence interval.

_____

8-5. **Female Heights** Exercise 5 in Section 8-3 of *Elementary Statistics* is based on the sample data listed below. Use STATDISK with a 0.05 significance level to determine whether there is sufficient evidence to support the claim that female statistics students exaggerate by reporting heights that are greater than their actual measured heights.

Reported Height	64	63	64	65	64	64	63	59	66	64
Measured Height	63.5	63.1	63.8	63.4	62.1	64.4	62.7	59.3	65.4	62.2

Test statistic:_____   Critical value(s):_____   *P*-value:_____

STATDISK conclusion about the null hypothesis:_____

STATDISK conclusion about the original claim:_____

Conclusion in your own words:_____

_____

Construct a 95% confidence interval estimate the mean difference between the reported and measured heights.

_____

Write a statement that *interprets* the confidence interval.

_____

8-6. **Crash Tests** Exercise 6 in Section 8-3 of *Elementary Statistics* includes sample data from low speed crash tests of five BMW cars. The repair costs were computed for a factory authorized repair center and an independent repair facility. The results are listed below.

Authorized Repair Center	$797	$571	$904	$1147	$418
Independent Repair Center	$523	$488	$875	$911	$297

Use STATDISK to determine whether there is sufficient evidence to support the claim that the independent center has lower repair costs. Use a 0.01 significance level.

Test statistic:_____   Critical value(s):_____   *P*-value:_____

STATDISK conclusion about the null hypothesis:_____

STATDISK conclusion about the original claim:_____

Conclusion in your own words:_____

_____

Construct a 95% confidence interval estimate the mean difference between the repair costs at the two factilites.

_____

8-7.   **Vaccine Testing**   Exercise 7 in Section 8-4 of *Elementary Statistics* notes that in a *USA Today* article about an experimental nasal spray vaccine for children, the following statement was presented: "In a trial involving 1602 children only 14 (1%) of the 1070 who received the vaccine developed the flu, compared with 95 (18%) of the 532 who got a placebo." The article also referred to a study claiming that the experimental nasal spray "cuts children's chances of getting the flu." Use STATDISK to determine whether there is sufficient sample evidence to support the stated claim.

Test statistic:_____   Critical value(s):_____   *P*-value:_____

STATDISK conclusion about the null hypothesis:_____

STATDISK conclusion about the original claim:_____

Conclusion in your own words:_____

_____

Construct a 95% confidence interval estimate the mean difference between the two population proportions.

_____

Write a statement that *interprets* the confidence interval.

_____

8-8.   **Testing Laboratory Gloves**   Exercise 14 in Section 8-4 of *Elementary Statistics* states that the *New York Times* ran an article about a study in which Professor Denise Korniewicz and other Johns Hopkins researchers subjected laboratory gloves to stress. Among 240 vinyl gloves, 63% leaked viruses. Among 240 latex gloves, 7% leaked viruses. At the 0.005 significance level, test the claim that vinyl gloves have a larger virus leak rate than latex gloves.

Test statistic:_____   Critical value(s):_____   *P*-value:_____

STATDISK conclusion about the null hypothesis:_____

STATDISK conclusion about the original claim:_____

Conclusion in your own words:_____

_____

Construct a 95% confidence interval estimate the mean difference between the two population proportions.

_____

Write a statement that *interprets* the confidence interval.

_____

8-9.   **Aluminum Can Manufacturing**   Exercise 4 in Section 8-5 of *Elementary Statistics* refers to the axial loads of aluminum cans listed in Data Set 12 in Appendix B. (An axial load is the maximum weight that the sides can support. It is measured by using a plate to apply increasing pressure to the top of the can until it collapses.) The sample of

0.0109 in. cans has axial loads with a mean of 267.1 lb and a standard deviation of 22.1lb. The sample of 0.0111 in. cans has axial loads with a mean of 281.8 lb and a standard deviation of 27.8 lb. Use a 0.05 significance level to test the claim that the samples come from populations with the same standard deviation.

Test statistic:_____ Critical value(s):_____ *P*-value:_____

STATDISK conclusion about the null hypothesis:_____

STATDISK conclusion about the original claim:_____

Conclusion in your own words:_____

_____

8-10. **Weight and Age** Exercise 6 in Section 8-5 of *Elementary Statistics* refers to data from the National Health Survey. For 804 men aged 25 to 34, the mean is 176 lb and the standard deviation is 35.0 lb. For 1657 men aged 65 to 74, the mean and standard deviation are 164 lb and 27.0 lb, respectively. At the 0.01 significance level, test the claim that the older men come from a population with a standard deviation less than that for men in the 25 to 34 age bracket.

Test statistic:_____ Critical value(s):_____ *P*-value:_____

STATDISK conclusion about the null hypothesis:_____

STATDISK conclusion about the original claim:_____

Conclusion in your own words:_____

_____

8-11. **Heights of Men and Women** Exercises in *Elementary Statistics* use data from randomly selected samples of men and women. When their heights are accurately measured, the results (in inches) are listed and summarized below (based on data from the National Health Survey).

Men: 73.0 71.6 68.7 65.9 70.4   → $\bar{x} = 69.920$, $s = 2.747$

Women: 63.9 67.1 63.2 64.7   → $\bar{x} = 64.725$, $s = 1.698$

Assuming that the heights of men and women have *equal* variances, use a 0.01 significance level to test the claim that men and women have the same mean height.

Test statistic:_____ Critical value(s):_____ *P*-value:_____

STATDISK conclusion about the null hypothesis:_____

STATDISK conclusion about the original claim:_____

continued

Conclusion in your own words:_____

_____

Assuming that the heights of men and women have *unequal* variances, use a 0.01 significance level to test the claim that men and women have the same mean height.

Test statistic:_____   Critical value(s):_____   *P*-value:_____

STATDISK conclusion about the null hypothesis:_____

STATDISK conclusion about the original claim:_____

Conclusion in your own words:_____

_____

Compare the two sets of results obtained above. Do both procedures lead to the same result?

_____

_____

Assuming that heights of men and women have *equal* variances, construct a 99% confidence interval estimate of the difference $\mu_1 - \mu_2$. Do the confidence interval limits contain 0, and what does this suggest about equality of the two population means?

_____

_____

Assuming that heights of men a women have *unequal* variances, construct a 99% confidence interval estimate of the difference $\mu_1 - \mu_2$. Do the confidence interval limits contain 0, and what does this suggest about equality of the two population means?

_____

_____

Compare the above two confidence intervals. Does the assumption of equal variances or unequal variances have much of an effect on the confidence interval limits?

_____

_____

8-12.   **Brain Right Cordate and Psychiatric Disorders** Exercises 7 and 8 in Section 8-6 of *Elementary Statistics* refer to the sample data summarized below, which are volumes (in mL) of the right cordate of the brain. Use STATDISK with a 0.01 significance level to test the claim that obsessive-compulsive patients and healthy persons have the same mean brain volume. Enter the results on the following page.

Obsessive-compulsive patients: $n = 10$, $\bar{x} = 0.34$, $s = 0.08$
Control group: $n = 10$, $\bar{x} = 0.45$, $s = 0.08$

Test statistic:_____   Critical value(s):_____   *P*-value:_____

STATDISK conclusion about the null hypothesis:_____

STATDISK conclusion about the original claim:_____

Conclusion in your own words:_____

_____

Construct a 99% confidence interval estimate of the difference between the mean brain volume for the patient group and the mean brain volume for the healthy control group. What does the confidence interval suggest about the difference between the two population means?

_____

_____

8-13. **Transformation Effects** Captopril is a drug designed to lower systolic blood pressure. When subjects were tested with this drug, their systolic blood pressure readings (in mm of mercury) were measured before and after the drug was taken, with the results given below (based on data from "Essential Hypertension: Effect of an Oral Inhibitor of Angiotensin-Converting Enzyme" by MacGregor et al, *British Medical Journal*, Volume 2).

Subject	A	B	C	D	E	F	G	H	I	J	K	L
Before	200	174	198	170	179	182	193	209	185	155	169	210
After	191	170	177	167	159	151	176	183	159	145	146	177

Use a 0.01 significance level to test the claim that captopril is effective in lowering systolic blood pressure. Enter the STATDISK results here.

Test statistic:_____   Critical value(s):_____   *P*-value:_____

STATDISK conclusion about the null hypothesis:_____

STATDISK conclusion about the original claim:_____

Conclusion in your own words:_____

_____

Construct a 99% confidence interval for the mean difference between the before and after readings.

_____

If 100 is added to every sample value, how are the hypothesis test results affected?

_____

_____

continued

If 100 is added to every sample value, how is the confidence interval affected?

_____

If every sample value is multiplied by 100, how are the hypothesis test results affected?
_____

In general, when working with matched pairs of sample data, how are the results affected if the same constant is added to every sample value?

_____

In general, when working with matched pairs of sample data, how are the hypothesis test results affected if every sample value is multiplied by the same positive constant?

_____

In general, when working with matched pairs of sample data, how are confidence intervals affected if every sample value is multiplied by the same positive constant?

_____

Based on the results from this experiment, how are hypothesis testing conclusions affected by the *scale* of values used?

_____

8-14.   **Outlier Effects**   Experiment 8-13 is designed to reveal effects of transformations of data. Design and conduct an experiment to determine the effect of an *outlier* on hypothesis test results and confidence intervals for matched pairs of sample data. What are the effects?

_____

_____

8-15.   **Using the Wrong Test**   Refer to the matched sample data in Experiment 8-13. Assume (incorrectly) that the two sets of sample values are from *independent* populations. How are the hypothesis test results affected?

_____

_____

Assuming (incorrectly) that the two sets of sample values are from *independent* populations, construct a 95% confidence interval. How is the confidence interval affected by assuming that the samples are from independent populations instead of populations of matched pairs?

_____

_____

8-16. Researchers at the Precision Instrument Corporation experiment with two production methods. The first method produces 10 defects among 50 sample items, while the second method produces 20 defects among 75 sample items. At the 0.05 level of significance, test the claim of equal proportions.

Test statistic:_____ Critical value(s):_____ *P*-value:_____

STATDISK conclusion about the null hypothesis:_____

STATDISK conclusion about the original claim:_____

Conclusion in your own words:_____

_____

Construct a 95% confidence interval for the difference between the two proportions of defective items.

_____

Repeat the hypothesis test and confidence interval construction after making these changes: Change 10/50 to 100/500 and change 20/75 to 200/750. (Note that the proportions are the same, but the samples are larger.)

How are the hypothesis test results affected?

_____

How is the confidence interval affected?

_____

In general, what do you conclude about the effect of increasing the sample size, while maintaining the same proportions?

_____

8-17. Refer to Data Set 11 in Appendix B of the textbook. (The data are already stored with the STATDISK program. The STATDISK file names are DURATION, INTERVAL, and HEIGHT.) What is fundamentally *wrong* with testing the claim that the intervals between eruptions have a mean equal to the mean height of the eruptions? Can STATDISK and other programs provide results for such a claim, or are there safeguards protecting you from such an error?

_____

_____

Instead of testing for equality between the two population means, is there any other characteristic that might be of interest to geologists. If so, describe that characteristic.

_____

9

Correlation
and
Regression

9-1   Linear Correlation and Regression

Sections 9-2, 9-3, and 9-4 in *Elementary Statistics* introduce the basic concepts of linear correlation and regression. The basic objective is to use sample paired data to determine whether there is a relationship between two variables and, if so, identify what the relationship is. Sections 9-2, 9-3, and 9-4 use the paired sample data in Table 9-1, reproduced below. We want to determine whether there is a relationship between the amount of a restaurant bill and the amount of the tip. If such a relationship exists, we want to identify it with an equation so that we can identify the rule that people follow when they leave tips. The STATDISK procedure follows.

Table 9-1   Paired Data for Six Dining Parties

Bill (dollars)	33.46	50.68	87.92	98.84	63.60	107.34
Tip (dollars)	5.50	5.00	8.08	17.00	12.00	16.00

STATDISK Procedure for Correlation and Regression

1.   Select **Analysis** from the main menu.

2.   Select the menu item of **Correlation and Regression**.

3.   Select a significance level, such as 0.05 or 0.01.

4.   Proceed to enter the data in the two columns. If there aren't too many pairs of data, it will be easy to enter them directly into the window. If there are many pairs of data, it is better to use the Sample Editor to enter and save the two data sets, then use Copy/Paste to copy them to the Correlation and Regression module.

5.   Click the **Evaluate** bar to get the correlation/regression results.

6.   Click **Plot 1** to get a scatter diagram with the regression line included.

7.   Click **Plot 2** to get a graph that includes the t test statistic and critical values.

For example, if you follow the above steps using the sample data in Table 9-1, the STATDISK results will be as shown on the following page, except for the result of Plot 2 which is not shown. The results include the linear correlation coefficient of $r = 0.82816$, the critical values of $r = \pm 0.81140$, and the conclusion that there is sufficient evidence to reject the null hypothesis (of no correlation) and support a claim of a linear correlation between the two variables. Also included are the y-intercept b_0 and slope b_1 of the estimated regression line. Using the STATDISK results, the estimated regression equation is

$$\hat{y} = -0.34728 + 0.14861x.$$

Section 9-4 of the textbook discusses values of total variation, explained variation, unexplained variation, standard error s_e, and coefficient of determination. The graph of the scatter diagram includes the regression line, as shown on the following page.

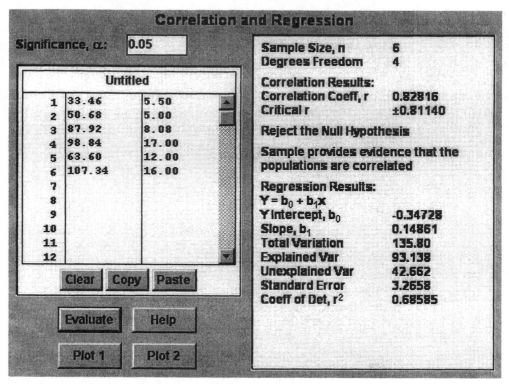

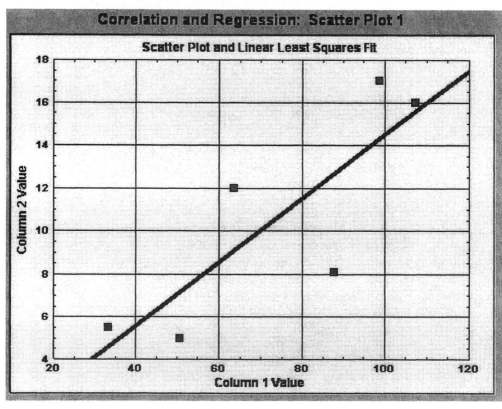

9-2 Multiple Regression

Section 9-5 of *Elementary Statistics* discusses multiple regression, and STATDISK does allow you to obtain multiple regression results. Once a collection of sample data has been entered, you can easily experiment with different combinations of columns to find the combination that is best. Here is the STATDISK procedure.

STATDISK Procedure for Multiple Regression

1. Select **Analysis** from the main menu.

2. Select **Multiple Regression** from the menu.

3. Enter the data in the different columns. Manually enter the data if the data set is not very large. If the data set is large, use Sample Editor to enter and save each individual column of data, then use Copy/Paste to copy the columns into the Multiple Regression module. For example, enter the first column in the Sample Editor module, save the data, then use Copy/Paste to copy that column in the Multiple Regression module. *Without closing the Multiple Regression window*, click on Data and Sample Editor and proceed to enter the second column of data, which you can then copy to the Multiple Regression module. Repeat this procedure until all columns are included.

4. After all columns of data are included, click on **Evaluate**.

5. You will now see a dialog box that allows you to select which columns are to be included, and which column is to be used for the dependent (*y*) variable. Indicate your choices. *Important*: When indicating the columns to be used, be sure to include the column corresponding to the dependent variable. See the sample display shown below. The dialog box shows that columns 1, 3, and 6 are to be included, and column 1 is the dependent variable.

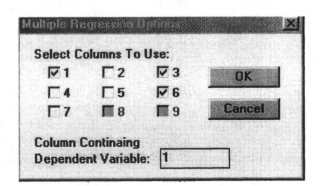

6. Click **OK** to obtain the results.

7. To use a different combination of variables, simply click the Evaluate bar. There is no need to enter the data again, because the columns continue to be available.

As an example, Table 9-3 from Section 9-5 of the textbook is reproduced below. If you enter the sample data included in Table 9-3 and select columns 1, 3, and 6 (with column 1 used for the dependent variable), the STATDISK results will be as shown below the table. The results correspond to this multiple regression equation:

$$\hat{y} = -374.30 + 18.820x_3 + 5.8748x_6$$

The results also include the adjusted coefficient of determination: Adjusted $R^2 = 0.75918$, as well as some other results.

TABLE 9-3   Data from Anesthetized Male Bears

Variable	Name	Sample Data							
y	WEIGHT	80	344	416	348	262	360	332	34
x_2	AGE	19	55	81	115	56	51	68	8
x_3	HEADLEN	11.0	16.5	15.5	17.0	15.0	13.5	16.0	9.0
x_4	HEADWDTH	5.5	9.0	8.0	10.0	7.5	8.0	9.0	4.5
x_5	NECK	16.0	28.0	31.0	31.5	26.5	27.0	29.0	13.0
x_6	LENGTH	53.0	67.5	72.0	72.0	73.5	68.5	73.0	37.0
x_7	CHEST	26	45	54	49	41	49	44	19

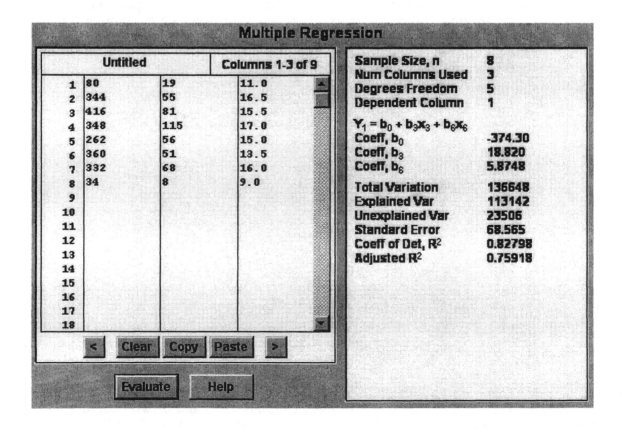

9-3 Modeling

Section 9-6 in *Elementary Statistics* discusses mathematical modeling. The objective is to find a mathematical function that "fits" or describes real-world data. Among the models discussed in the textbook, we will describe how STATDISK can be used for the linear, quadratic, logarithmic, exponential, and power models.

To illustrate the use of STATDISK, consider the sample data in Table 9-5 from the textbook, reproduced below. As in the textbook, we will use the coded year values for x, so that $x = 1, 2, 3, ..., 10$. The y values are the populations (in millions) of 5, 10, 17, ..., 227.

Table 9-5 Population (in millions) of the United States

Year	1800	1820	1840	1860	1880	1900	1920	1940	1960	1980
Coded Year	1	2	3	4	5	6	7	8	9	10
Population	5	10	17	31	50	76	106	132	179	227

Linear Model: $y = a + bx$

The linear model can be obtained by using STATDISK's correlation and regression module. The procedure is described in Section 9-1 of this manual/workbook. Select **Analysis**, then **Correlation and Regression**. Enter the values of x in the first column, enter the values of y in the second column, then click **Evaluate**. For the data in Table 9-5, enter the coded year values of 1, 2, 3, ..., 10 for the first column and enter the population values of 5, 10, 17, ..., 227 for the second column. The result will be as shown below.

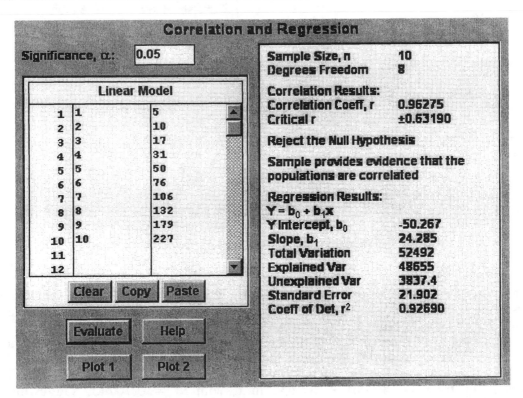

The above STATDISK display describes key results for the linear model. The resulting function is $y = -50.267 + 24.285x$, and the coefficient of determination is given as $r^2 = 0.92690$. The high value of r^2 suggests that the linear model is a reasonably good fit.

Quadratic Model: $y = ax^2 + bx + c$

Select **Analysis**, then **Multiple Regression**. In the dialog box, enter the values of x in the first column, enter the values of x^2 in the second column, and enter the values of y in the third column. When indicating the columns to be used, leave check marks next to 1, 2, and 3; enter 3 for the column representing the dependent variable.

The display shown below corresponds to the quadratic model used with the sample data in Table 9-5. Note that there are three columns of data representing x, x^2, and y. The results show that the function has the form given as $y = 8.5667 - 5.1318x + 2.6742x^2$. Remember, the coefficient b_1 corresponds to x and b_2 corresponds to x^2. The coefficient of determination is $R^2 = 0.99883$, suggesting a better fit than the linear model.

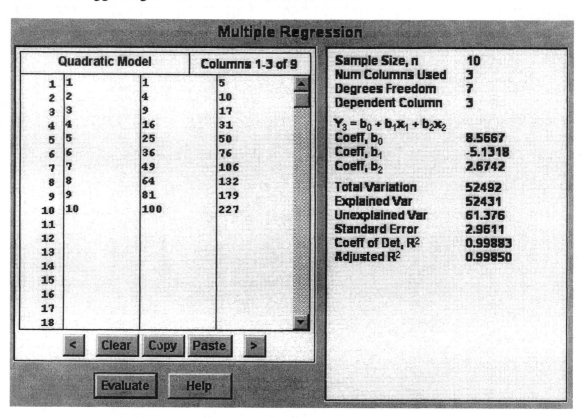

Logarithmic Model: $y = a + b \ln x$

Select **Analysis**, then **Correlation and Regression**. In the dialog box, enter the values of $\ln x$ in the first column and enter the values of y in the second column. (The values of $\ln x$ can be found using STATDISK's Sample Transformations module.)

The display shown on the top of the next page results from the logarithmic model used with the sample data in Table 9-5. Note that the first column consists of values of $\ln 1$, $\ln 2$, $\ln 3$, ..., $\ln 10$. (These values can be obtained by using the Sample Transformation module.) The function is given by $y = -49.061 + 87.631 \ln x$, with $R^2 = 0.70745$, suggesting that this model does not fit as well as the linear or quadratic models. Of the three models considered so far, the quadratic model appears to be best.

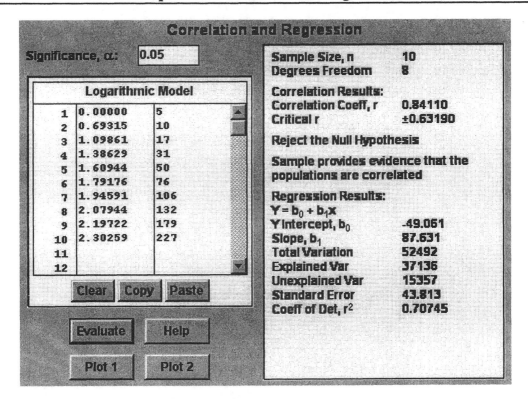

Exponential Model: $y = ab^x$

The exponential model is tricky, but it can be obtained using STATDISK. Select **Analysis**, then **Correlation and Regression**. Enter the values of x in the first column, and enter the values of ln y in the second column. (The values of ln y can be found using STATDISK's Sample Transformations module.) When you get the results from STATDISK, the value of the coefficient of determination is correct, but the values of a and b in the exponential model must be computed as follows:

To find the value of a: Evaluate e^{b0} where b_0 is given by STATDISK.

To find the value of b: Evaluate e^{b1} where b_1 is given by STATDISK.

Using the data in Table 9-5, this procedure results in the STATDISK display on the top of the next page. The value of $R^2 = 0.96818$ is OK as is, but the values of a and b must be computed from the STATDISK results as shown below:

$a = e^{b0} = e^{1.5618} = 4.7674$

$b = e^{b1} = e^{0.41751} = 1.5182$

Using these values of a and b, we express the exponential model as

$$y = 4.7674(1.5182^x)$$

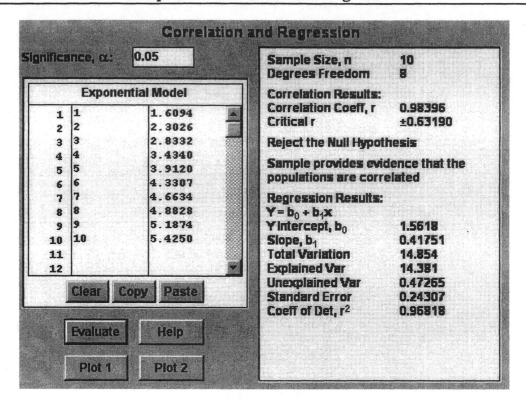

Power Model: $y = ax^b$

The power model is also tricky, but it can be obtained using STATDISK. Select **Analysis**, then **Correlation and Regression**. Enter the values of ln x in the first column, and enter the values of ln y in the second column. (The values of ln x and ln y can be found using STATDISK's Sample Transformations module.) When you get the results from STATDISK, the value of the coefficient of determination is correct, but the values of a and b in the power model are found as follows:

To find the value of a: Evaluate e^{b0} where b_0 is given by STATDISK.

The value of b is the same as the value of b_1 given by STATDISK.

Using the data in Table 9-5, this procedure results in the STATDISK display on the top of the next page. The value of $R^2 = 0.97502$ is OK as is, and it suggests that the power model is not as good as the quadratic model. The values of a and b are found from the STATDISK results as shown below:

$a = e^{b0} = e^{1.2442} = 3.4702$

$b = b_1$ from STATDISK $= 1.7305$

Using these values of a and b, we express the power model as

$$y = 3.4702(x^{1.7305})$$

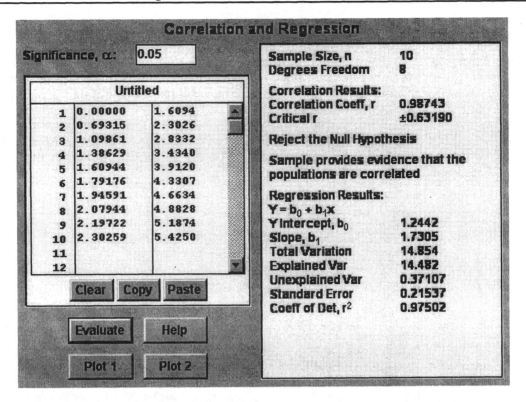

The rationale underlying the methods for the exponential and power models is based on transformations of equations. In the exponential model of $y = ab^x$, for example, taking natural logarithms of both sides yields $\ln y = \ln a + x (\ln b)$, which is the equation of a straight line. STATDISK can be used to find the equation of this straight line that fits the data best; the intercept will be $\ln a$ and the slope will be $\ln b$, but we need the values of a and b, so we solve for them as described above. Similar reasoning is used with the power model. We will not consider the use of STATDISK for the logistic model of $y = \dfrac{c}{1 + ae^{-bx}}$.

CHAPTER 9 EXPERIMENTS: Correlation and Regression

9-1.   **Bear Weights and Ages** Refer to the Table 9-3 data from anesthetized male bears included in the textbook and reproduced in Section 9-2 of this manual/workbook. Using the Correlation and Regression module, enter the values for WEIGHT in column 1 and enter the values for AGE in column 2.

   a.   Display the scatter diagram of the paired WEIGHT/AGE data. Based on that scatter diagram, does there appear to be a relationship between the weights of bears and their ages? If so, what is it?

   _____

   b.   Find the value of the linear correlation coefficient r._____

   c.   Assuming a 0.05 level of significance, what do you conclude about the correlation between weights and ages of bears?

   _____

   d.   Find the equation of the regression line. (Use WEIGHT as the x predictor variable, and use AGE as the y response variable.) _____

   e.   What is the best predicted age of a bear that weighs 300 lb? (*Caution*: See the "Prediction" subsection in Section 9-3 of the textbook.) _____

9-2.   **Effect of Transforming Data** The ages used in Experiment 9-1 are in months. Convert them to days by multiplying each age by 30, then repeat Experiment 9-1 and enter the responses here:

   a.   Display the scatter diagram of the paired WEIGHT/AGE data. Based on that scatter diagram, does there appear to be a relationship between the weights of bears and their ages? If so, what is it?

   _____

   b.   Find the value of the linear correlation coefficient r._____

   c.   Assuming a 0.05 level of significance, what do you conclude about the correlation between weights and ages of bears?

   _____

   d.   Find the equation of the regression line. (Use WEIGHT as the x predictor variable, and use AGE as the y response variable.) _____

   e.   What is the best predicted age of a bear that weighs 300 lb? (*Caution*: See the "Prediction" subsection in Section 9-3 of the textbook.) _____

   f.   After comparing the responses obtained in Experiment 9-1 to those obtained here, describe the general affect of changing the scale for one of the variables.

   _____

9-3.   **Effect of More Data**   Experiment 9-1 used the sample data in Table 9-3, which summarizes measurements for only eight bears. Using File and Open, retrieve the weights of all 54 bears that are included in Data Set 7 in Appendix B of the textbook. (That data set is already available with the name BEARWT.) Use Copy/Paste to copy those weights to the Correlation and Regression module. Next, retrieve the 54 ages (stored with the file name of BEARAGE) and copy them to the Correlation and Regression module (column 2). Repeat Experiment 9-1 using the data from all 54 bears. Enter the responses to parts (a) through (e) in the spaces below.

   a.   Display the scatter diagram of the paired WEIGHT/AGE data. Based on that scatter diagram, does there appear to be a relationship between the weights of bears and their ages? If so, what is it?

   _____

   b.   Find the value of the linear correlation coefficient r. _____

   c.   Assuming a 0.05 level of significance, what do you conclude about the correlation between weights and ages of bears?

   _____

   d.   Find the equation of the regression line. (Use WEIGHT as the x predictor variable, and use AGE as the y response variable.) _____

   e.   What is the best predicted age of a bear that weighs 300 lb? (*Caution*: See the "Prediction" subsection in Section 9-3 of the textbook.) _____

   f.   After comparing the responses obtained in Experiment 9-1 to those obtained here, what do you conclude about the changes that occur when we use the sample of 54 bears instead of the smaller sample of 8 bears?

   _____

9-4.   **Effect of No Variation for a Variable**   Use STATDISK with the following paired data and obtain the indicated results.

x	1	2	3	4	5	7	7	9
y	5	5	5	5	5	5	5	5

   a.   Based on an examination of the displayed scatter diagram, does there appear to be a relationship between x and y? If so, what is it?

   _____

   b.   What happens when you try to find the value of r? Why?

   _____

   c.   What do you conclude about the correlation between x and y? What is the equation of the regression line?

   _____

9-5.   **Garbage Data for Predicting Household Size**   Data Set 5 in Appendix B of
        Elementary Statistics consists of data from the Garbage Project at the University of
        Arizona. That data set is also included with the STATDISK program. (See Data Set 5
        in Appendix B, where the columns of data are identified with the STATDISK names of
        HHSIZE, METAL, PAPER, PLAS, GLASS, FOOD, YARD, TEXT, OTHER, and
        TOTAL. Use household size (HHSIZE) as the response y variable. For each given
        predictor x variable, find the value of the linear correlation coefficient, the equation of
        the regression line, and the value of the coefficient of determination r^2. Enter the
        results below.

	r	Equation of regression line	r^2
Metal	___	_____	___
Paper	___	_____	___
Plastic	___	_____	___
Glass	___	_____	___
Food	___	_____	___
Yard	___	_____	___
Text	___	_____	___
Other	___	_____	___
Total	___	_____	___

Based on the above results, which single independent variable appears to be the best
predictor of household size? Why?

9-6.   Use the same data set described in Experiment 9-5. Let household size (HHSIZE) be
        the dependent y variable and use the given predictor x variables to fill in the results
        below.

	Multiple regression eq.	R^2	Adj. R^2
Metal and Paper	_____	___	____
Plastic and Food	_____	___	____
Metal, Paper, Glass	_____	___	____
Metal, Paper, Plastic, Glass	_____	___	____

Based on the above results, which of the multiple regression equations appears to best
fit the data? Why?

9-7. **Diamond Data** Refer to Data Set 3 in Appendix B of *Elementary Statistics*, and use only the variables for price (DMDPRICE), carat (DMDCARAT), color (DMDCOLOR), and clarity (DMDCLRTY). Let the price be the dependent y variable and experiment with different combinations of independent variables to find the equation that best fits the sample data. Identify that equation and describe why it is best.

9-8. **Movie Data** Refer to Data Set 15 in Appendix B. Let the movie gross amount (MVGRS) be the dependent y variable and experiment with different combinations of budget amount (MVBUD), length (MVLEN), and viewer rating (MVRAT). Find the equation that best fits the sample data. Identify that equation and describe why it is best.

9-9. **Cigarette Data** Use Data Set 8 from Appendix B of the textbook, which is stored with the STATDISK program. The STATDISK file names are TAR, NICOTINE, and CO. Assume that we want to predict the amount of nicotine in a cigarette, based on the amount of tar and carbon monoxide. Use NICOTINE as the dependent variable and use TAR and/or CO (carbon monoxide) for independent variables. Find the equation that is best for predicting NICOTINE in a cigarette and describe why it is best.

9-10. **Old Faithful** Use Data Set 11 from Appendix B of the textbook. That data set is stored with the STATDISK files DURATION, INTERVAL, and GEYSERHT. Assume that we want to predict the time of the next eruption of Old Faithful, so we need to predict a value of the variable INTERVAL, given the duration and height of the last eruption. We therefore select INTERVAL as the dependent variable. What is the best equation for predicting the duration time to the next eruption? Why?

What is the best predicted time before the next eruption if the previous eruption lasted for 210 seconds and had a height of 275 ft? _____

9-11. **Shad Fish** Exercise 5 in Section 9-6 of *Elementary Statistics* includes the weights (in pounds) of shad fish harvested in the Hudson River (based on data from the New York State Department of Environmental Conservation). Those weights are reproduced in the table on the next page. Use STATDISK with the sample data to find the equations and coefficients of determination for the indicated models, and enter the results in the spaces that follow the table.

Year	Pounds
1980	1,313,100
1981	620,200
1982	378,900
1983	459,400
1984	701,400
1985	756,064
1986	798,768
1987	684,182
1988	782,932
1989	485,700
1990	463,529
1991	329,368
1992	265,598
1993	138,210
1994	157,672
1995	190,607
1996	135,629
1997	93,688

	Equation	R^2
Linear	_____	_____
Quadratic	_____	_____
Logarithmic	_____	_____
Exponential	_____	_____
Power	_____	_____

Based on the above results, which model appears to best fit the data? Why?

_____

What is the best predicted value for 1998?   _____

9-12. **Swimming Records** Exercise 6 in Section 9-6 of *Elementary Statistics* includes the given data for world record times in men's swimming. Use STATDISK with the sample data to find the equations and coefficients of determination for the indicated models. Enter the results on the next page.

Year	1912	1924	1957	1968	1972	1976	1988	1994
Time (sec)	61.6	57.4	54.6	52.2	51.22	49.99	48.42	48.21

	Equation	R^2
Linear	_____	_____
Quadratic	_____	_____
Logarithmic	_____	_____
Exponential	_____	_____
Power	_____	_____

Based on the above results, which model appears to best fit the data? Why?

_____

Is one of the above models considerably better than the others?

_____

9-13. **Return on Investment**   Exercise 7 in Section 9-6 of *Elementary Statistic* includes the investment/revenue amounts listed below.

Amount Invested (thousands of dollars)	1	2	5	11	20	31	41	46	48
Revenue (dollars)	2001	2639	3807	5219	6629	7899	8834	9250	9409

Of the linear, quadratic, logarithmic, exponential, and power models, which is best? Why?

_____

9-14. **Distance/Time for Dropped Golf Ball**   Exercise 8 in Section 9-6 of *Elementary Statistics* includes the following results from a physics experiment in which a golf ball is dropped from a tall building and the distances (in feet) below the point of release are recorded for different times (in seconds) that the ball has fallen.

Time	0	0.5	1.0	1.5	2.0	2.5	3.0	3.5	4.0	4.5	5.0
Distance	0	4.0	15	35	63	100	143	194	253	320	396

Of the linear, quadratic, logarithmic, exponential, and power models, which is best? Why?

_____

10

Multinomial Experiments and Contingency Tables

10-1 Multinomial Experiments

Section 10-2 of *Elementary Statistics* states that each data set in that section consists of qualitative data that have been separated into different categories. The main objective is to determine whether the distribution agrees with or "fits" some claimed distribution. Also, a multinomial experiment is defined as follows:

A **multinomial experiment** is an experiment that meets the following conditions.

1. The number of trials is fixed.

2. The trials are independent.

3. All outcomes of each trial must be classified into exactly one of several different categories.

4. The probabilities for the different categories remain constant for each trial.

STATDISK Procedure for Multinomial Experiments

1. Select **Analysis** from the main menu.

2. Select **Multinomial Experiments** from the submenu.

3. You now are presented with the following two options:

   - Equal Expected Frequencies
   - Unequal Expected Frequencies

   If you want to test the claim that the different categories are all equally likely, then select "Equal Expected Frequencies." If you want to test the claim that the different categories occur with some claimed proportions (not all equal), select the second item of "Unequal Expected Frequencies."

4. In the dialog box that now appears, select a significance level, such as 0.05 or 0.01.

5. Enter the observed frequencies. If you select "Unequal Expected Frequencies" in Step 3, you must also enter expected values. You can enter the actual expected frequencies, or you can enter the expected *proportions*. (If you choose to enter the expected proportions, indicate that choice by clicking on the circle to the left of the option identified "As proportions.") For example, if you want to test the claim that absences on the different weekdays are expected to occur with frequencies of 40, 10, 10, 10, 30, you can enter those expected frequencies. If you want to test the claim that absences are expected to occur with the proportions of 0.4, 0.1, 0.1, 0.1, and 0.3, select the case of "Unequal Expected Frequencies," then select the option of entering the expected frequencies "As proportions" and proceed to enter those proportions in the column for expected values.

6. Click on the **Evaluate** bar.

7.   Click on **Plot** to obtain a graph of the χ^2 distribution that includes the test statistic and critical value.

Section 10-2 in *Elementary Statistics* includes an example involving an analysis of the last digits of the reported distances of homeruns hit by Mark McGwire when he broke a major baseball record in 1998. The data from Table 10-2 are reproduced here.

Table 10-2   Last Digits of Mark McGwire's Homerun Distances

Last Digit	0	1	2	3	4	5	6	7	8	9
Frequency	55	2	1	1	0	3	0	2	4	2

As in the textbook, we want to text the claim that the digits do not all occur with the same frequency. Using the above STATDISK procedure, select **Analysis**, then **Multinomial Experiments**, then select the option of **Equal Expected Frequencies**. We proceed to enter the observed frequencies in the dialog box. Using a 0.05 significance level, we get the results shown below. We can see that the *P*-value of 0.0000 suggests that we reject the null hypothesis that the frequencies are the same. We support the claim that the last digits occur with different frequencies. This result suggests that the homerun distances were not actually measured; they were estimated. The top of the next page shows the graph displayed when clicking on the **Plot** button. That graph shows the relative placement of the test statistic and critical value.

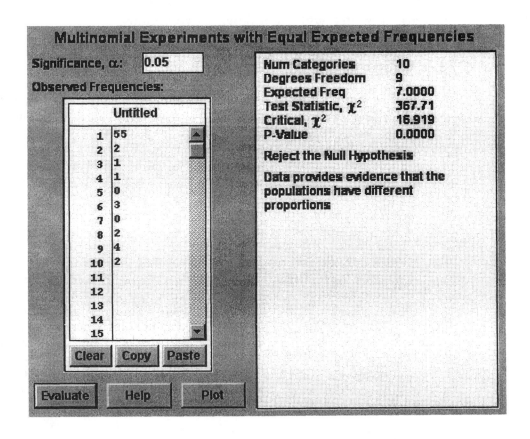

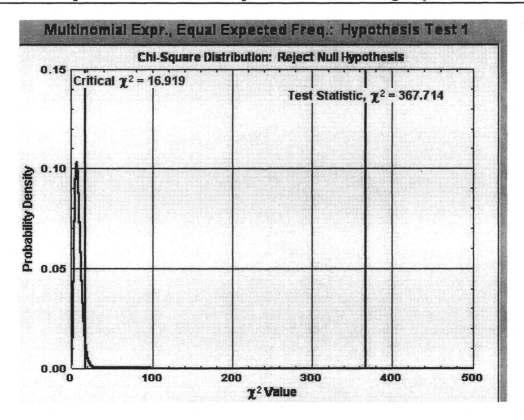

10-2 Contingency Tables

Section 10-3 of *Elementary Statistics* discusses contingency tables. To use STATDISK with a contingency table, use the following procedure.

STATDISK Procedure for Contingency Tables

1. Select **Analysis** from the main menu.

2. Select **Contingency Tables**.

3. In the dialog box that appears, enter a significance level such as 0.05 or 0.01.

4. Enter the contingency table in the data-entry field of the screen. Enter the frequencies as they appear in the table. For example, to enter the frequencies from a 2×3 contingency table, use two rows and three columns.

5. Click on the **Evaluate** bar.

6. Click on **Plot** to obtain a graph of the χ^2 distribution that includes the test statistic and critical value.

The example in Section 10-3 of the textbook uses the Titanic mortality data summarized in the contingency table of Table 10-1, which is reproduced on the next page.

Table 10-1   Titanic Mortality

	Men	Women	Boys	Girls	Total
Survived	332	318	29	27	**706**
Died	1360	104	35	18	**1517**
Total	**1692**	**422**	**64**	**45**	**2223**

We want to use a 0.05 significance level to test the claim that when the Titanic sank, whether someone survived or died is *independent* of whether the person is a man, woman, boy, or girl. The STATDISK results will be as shown below. See how the sample data are entered in the left portion of the dialog box. (The fourth column of 27 and 18 is not visible in this screen display, but that column of data has been entered.) Note that the observed frequencies are entered as they appear in Table 10-1, but the row and column totals are not entered.

We can see that the STATDISK display includes the important elements we need to make a decision. The *P*-value of 0.0000 indicates that we should reject the null hypothesis of independence between the row and column variables. The test statistic and critical value are also provided in the display. (There is only one critical value, because all hypothesis tests of this type are right-tailed.) It appears that whether a person survived the Titanic and whether that person is a man, woman, boy, or girl, are dependent variables.

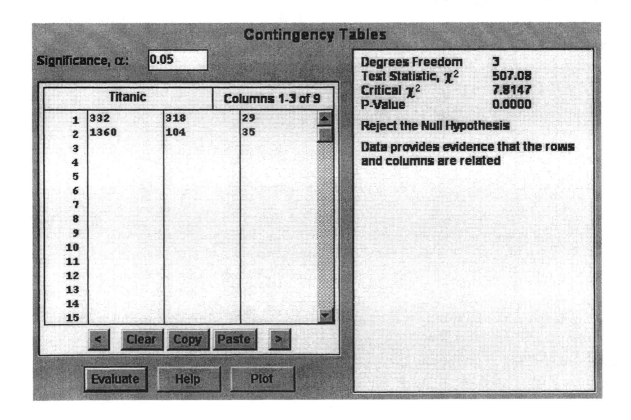

CHAPTER 10: Multinomial Experiments and Contingency Tables

10-1. **Flat Tire and Missed Class** Exercise 3 in Section 10-2 from *Elementary* Statistics refers to four car-pooling students who missed a test and gave an excuse of a flat tire. On the makeup test, the instructor asked the students to identify the particular tire that went flat. If they really didn't have a flat tire, would they be able to identify the same tire? The author asked students to identify the tire they would select. The results are listed in the following table (except for one student who selected the spare). Use a 0.05 level of significance to test the author's claim that the results fit a uniform distribution.

Tire	Left front	Right front	Left rear	Right rear
Number selected	11	15	8	6

Test statistic:_____   Critical value:_____   *P*-value:_____

Conclusion:_____

_____

What does the result suggest about the ability of students to select the same tire when they really didn't have a flat?

_____

10-2. **Do Car Crashes Occur on Different Days with the Same Frequency?** Exercise 5 in Section 10-2 from *Elementary Statistics* includes data from a sample of motor vehicle deaths in Montana. The numbers of fatalities for the different days of the week are listed below. At the 0.05 significance level, test the claim that accidents occur with equal frequency on the different days.

Day	Sun	Mon	Tues	Wed	Thurs	Fri	Sat
Number of fatalities	31	20	20	22	22	29	36

Test statistic:_____   Critical value:_____   *P*-value:_____

Conclusion:_____

_____

10-3. **Effect of Transformation** Repeat Experiment 10-2 after multiplying each observed frequency by 10. Enter the STATDISK results here.

Test statistic:_____   Critical value:_____   *P*-value:_____

Conclusion:_____

_____

In general, if each observed frequency is multiplied by the same integer greater than 1, how are the results affected?

_____

10-4.   **Testing for Uniform Distribution**   Exercise 6 in Section 10-2 of *Elementary Statistics* includes the following data: A study was made of 147 industrial accidents that required medical attention. Among those accidents, 31 occurred on Monday, 42 on Tuesday, 18 on Wednesday, 25 on Thursday, and 31 on Friday (based on results from "Counted Data CUSUM's," by Lucas, *Technometrics,* Vol. 27, No. 2). Test the claim that accidents occur with equal proportions on the five workdays.

Test statistic:_____   Critical value:_____   *P*-value:_____

Conclusion:_____

_____

10-5.   **Do Industrial Accidents Fit the Claimed Distribution?**   Use a 0.05 significance level and the data from the preceding experiment to test the claim of a safety expert that accidents are distributed on workdays as follows: 30% on Monday, 15% on Tuesday, 15% on Wednesday, 20% on Thursday, and 20% on Friday.

Test statistic:_____   Critical value:_____   *P*-value:_____

Conclusion:_____

_____

10-6.   **Old Faithful**   Refer to the time intervals between eruptions of the Old Faithful geyser, as listed in Data Set 11 of Appendix B in the textbook. That data set is stored with the file name of INTERVAL. Test the claim that the time intervals are uniformly distributed among the five categories of 55-64, 65-74, 75-84, 85-94, and 95-104.

Test statistic:_____   Critical value:_____   *P*-value:_____

Conclusion:_____

_____

10-7.   **Testing the Random Number Generator**   Use STATDISK to randomly generate 1000 digits between 0 and 9 inclusive, and enter the results below. (To generate the digits, select **Data**, then **Uniform Generator**, and enter 1000 for *n*, a minimum of 0, a maximum of 9, and 0 decimal places.) Use STATDISK to obtain a histogram so that the frequency counts for the different classes can be identified. Enter the frequencies below, then use them to test the claim that the digits occur with the same frequency.

Digit	0	1	2	3	4	5	6	7	8	9
Frequency										

Test statistic:_____   Critical value:_____   *P*-value:_____

Conclusion:_____

_____

10-8. **Testing a Normal Distribution** In this experiment we will use STATDISK's ability to generate normally distributed random numbers. We will then test the sample data to determine if they actually do fit a normal distribution.

a. Generate 1000 random numbers from a normal distribution with a mean of 100 and a standard deviation of 15. (IQ scores have these parameters.) Select **Data**, then **Normal Generator**.

b. Arrange the generated data in order. Use **Copy/Paste** to copy the data to the **Sample Editor** module, where the **Format** option allows you to sort the data so that they are arranged in order.

c. Examine the sorted list and determine the frequency for each of the categories listed below. Enter those frequencies in the spaces provided. (The expected frequencies were found using the methods of Chapter 5 in the textbook.)

	Observed Frequency	Expected Frequency
Below 55:	_____	1
55-70:	_____	22
70-85:	_____	136
85-100:	_____	341
100-115:	_____	341
115-130:	_____	136
130-145:	_____	22
Above 145:	_____	1

d. Use STATDISK to test the claim that the randomly generated numbers actually do fit a normal distribution with mean 100 and standard deviation 15.

Test statistic:_____ Critical value:_____ P-value:_____

Conclusion:_____

_____

10-9. **Testing for Independence Between Early Discharge and Rehospitalization of Newborn** Exercise 4 in Section 10-3 of *Elementary Statistics* includes the contingency table reproduced on the top of the next page. Use a 0.05 significance level to test the claim that whether the newborn was discharged early or late is independent of whether the newborn was rehospitalized within a week of discharge. Enter the results on the following page.

	Rehospitalized within a week of discharge?	
	Yes	No
Early discharge (less than 30 hours)	622	3997
Late discharge (30-78 hours)	631	4660

Test statistic:_____   Critical value:_____   _P_-value:_____

Conclusion:_____

_____

10-10. **Testing for Discrimination**   Exercise 6 in Section 10-3 of the textbook notes that in the judicial case _United States v. City of Chicago,_ fair employment practices were challenged. A minority group (group A) and a majority group (group B) took the Fire Captain Examination. Assume that the study began with predetermined sample sizes of 24 minority candidates (Group A) and 562 majority candidates (Group B), with the results as shown in the table. At the 0.05 significance level, test the claim that the proportion of minority candidates who pass is the same as the proportion of majority candidates who pass.

	Pass	Fail
Group A	10	14
Group B	417	145

Test statistic:_____   Critical value:_____   _P_-value:_____

Conclusion:_____

_____

10-11. **Is Scanner Accuracy the Same for Specials?**   Exercise 8 in Section 10-3 from the textbook describes a study of store checkout scanning systems. Samples of purchases were used to compare the scanned prices to the posted prices. The accompanying table summarizes results for a sample of 819 items. When stores use scanners to check out items, we want to determine whether the error rates are the same for regular-priced items as they are for advertised-special items.

	Regular-Priced Items	Advertised-Special Items
Undercharge	20	7
Overcharge	15	29
Correct price	384	364

Test statistic:_____   Critical value:_____   _P_-value:_____

Conclusion:_____

_____

10-12. **Survey Refusals and Age Bracket**   Exercise 16 in Section 10-3 of *Elementary Statistics* cites a study of people who refused to answer survey questions. Results are summarized in the accompanying table. At the 0.01 significance level, test the claim that the cooperation of the subject (response, refusal) is independent of the age category.

	Age					
	18–21	22–29	30–39	40–49	50–59	60 and over
Responded	73	255	245	136	138	202
Refused	11	20	33	16	27	49

Test statistic:_____   Critical value:_____   *P*-value:_____

Conclusion:_____

_____

What results change if the table is transposed by interchanging the rows and columns?

_____

How are the results affected if the order of the rows is switched?

_____

How are the results affected by the presence of an outlier? If we change the first entry from 73 to 7300, are the results dramatically affected?

_____

How are the results affected if one or more cells has an expected frequency less than 5?

_____

11

Analysis of Variance

11-1   One-Way Analysis of Variance

11-1 One-Way Analysis of Variance

Section 11-2 in *Elementary Statistics* discusses one-way analysis of variance, and STATDISK is programmed for the methods described in that section. STATDISK is not programmed for two-way analysis of variance as described in Section 11-3 of the textbook. Therefore, two-way analysis of variance will not be considered in this manual/workbook.

In describing procedures for one-way analysis of variance in Section 11-2 of the textbook, samples of the same size are considered along with samples of different sizes. STATDISK will work with both types; the samples need not have the same number of values.

In one-way analysis of variance, the term "one-way" is used because the sample data are separated into groups according to one characteristic or "factor". For example, Section 11-2 discusses the data in Table 11-1, reproduced below. Specifically, Section 11-2 deals with the head injury measurements for the four categories of cars (subcompact, compact, midsize, and full size). In this case, the factor is the car category. That is, the head injury measurements are analyzed in the context of the factor consisting of four separate car categories.

Table 11-1 **Injuries to Car Crash Test Dummies**

	Head Injury (hic)	Chest Deceleration (g)	Left Femur Load (lb)
Subcompact Cars			
Ford Escort	681	55	595
Honda Civic	428	47	1063
Hyundai Accent	917	59	885
Nissan Sentra	898	49	519
Saturn SL4	420	42	422
Compact Cars			
Chevrolet Cavalier	643	57	1051
Dodge Neon	655	57	1193
Mazda 626 DX	442	46	946
Pontiac Sunfire	514	54	984
Subaru Legacy	525	51	584
Midsize Cars			
Chevrolet Camaro	469	45	629
Dodge Intrepid	727	53	1686
Ford Mustang	525	49	880
Honda Accord	454	51	181
Volvo S70	259	46	645
Full Size Cars			
Audi A8	384	44	1085
Cadillac Deville	656	45	971
Ford Crown Victoria	602	39	996
Oldsmobile Aurora	687	58	804
Pontiac Bonneville	360	44	1376

The following procedure describes how STATDISK can be used with such collections of sample data to test the claim that the different samples come from populations with the same mean. For the head injury data of Table 11-1, the claim of equal means leads to these hypotheses:

H_0: $\mu_1 = \mu_2 = \mu_3 = \mu_4$

H_1: At least one of the four population means is different from the others.

STATDISK Procedure for One-Way Analysis of Variance

1.   Select **Analysis** from the main menu.

2.   Select **One-Way Analysis of Variance** from the submenu.

3.   In the dialog box, enter a significance level, such as 0.05 or 0.01.

4.   Enter the data in the individual columns. If the sample sizes are all small, manually enter the data. If there are many values, strongly consider using the Sample Editor for entering and saving the data in individual files, then use Copy/Paste to copy the data to the One-Way Analysis of Variance module.

5.   Click on the **Evaluate** bar.

6.   Click on the **Plot** bar to obtain a graph that includes the critical value and test statistic.

If you use the above steps with the data in Table 11-1, the STATDISK result will appear as shown on the top of the next page. The STATDISK display shows only the first four columns of sample data, but the fourth column has been entered and it can be seen by clicking on the > button.

The *P*-value of 0.42157 indicates that there is not sufficient sample evidence to warrant rejection of the null hypothesis that $\mu_1 = \mu_2 = \mu_3 = \mu_4$. The test statistic of $F = 0.99217$ is also provided along with the critical value of $F = 3.2389$. The values of the SS and MS components are also provided.

If you click on the **Plot** button, you will get the graph shown on the following page. The graph, which includes the test statistic and critical value, is easier to understand when it is viewed in color on a computer monitor.

Caution: It is easy to feed STATDISK data that can be processed quickly and painlessly, but we should *think* about what we are doing. We should consider the assumptions for the test being used, and we should *explore* the data before jumping into a formal procedure such as analysis of variance. Carefully examine the boxplots and graph of means that accompany the Chapter Problem for Chapter 11 in the textbook. In general, consider exploring the important characteristics of data, including the center (through means and medians), variation (through standard deviations and ranges), distribution (through histograms and boxplots), outliers, and any changing patterns over time.

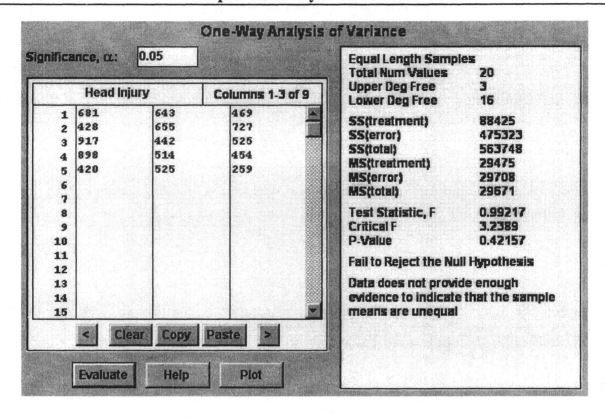

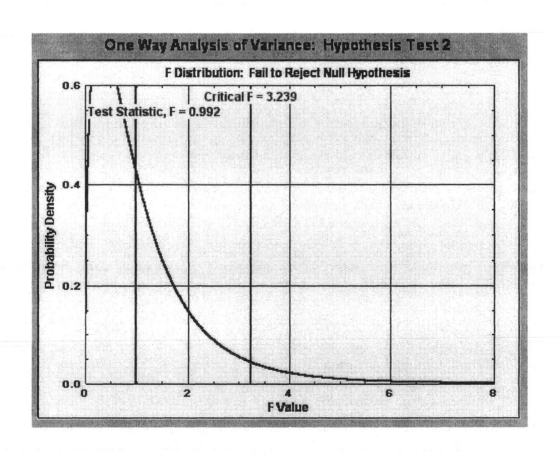

CHAPTER 11 EXPERIMENTS: Analysis of Variance

11-1. **Chest Deceleration Measurements**   The example in Section 11-1 of this manual/ workbook used the head injury data from Table 11-1. Use STATDISK with a 0.05 significance level to test the null hypothesis that the different weight categories have the same mean chest deceleration value. Enter the results below.

SS(treatment): _____   MS(treatment): _____   Test statistic F: _____

SS(error):   _____   MS(error):   _____ P-value: _____

SS(total):   _____

Conclusion:_____

_____

11-2. **Left Femur Load Measurements**   The example in Section 11-1 of this manual/ workbook used the head injury data from Table 11-1. Use STATDISK with a 0.05 significance level to test the null hypothesis that the different weight categories have the same mean left femur load. Enter the results below.

SS(treatment): _____   MS(treatment): _____   Test statistic F: _____

SS(error):   _____   MS(error):   _____ P-value: _____

SS(total):   _____

Conclusion:_____

_____

11-3. **Effects of Transformations and Outliers**   After completing Experiment 11-2, you can modify the data set by adding a constant or by multiplying by a constant. Make such modifications to answer the following questions.

How are the analysis of variance results affected if the same positive constant is added to every sample value?

_____

How are the analysis of variance results affected if every sample value is multiplied by the same positive constant?

_____

How are the analysis of variance results affected if the values of one of the samples are each increased by a large constant?

_____

How are the results affected if one sample value is changed to become an outlier?

_____

11-4. **Old Faithful** Listed below are time intervals (in minutes) between eruptions of the Old Faithful geyser in Yellowstone National Park. Test the null hypothesis that the four samples come from populations with the same mean.

1951	1985	1995	1996
74	89	86	88
60	90	86	86
74	60	62	85
42	65	104	89
74	82	62	83
52	84	95	85
65	54	79	91
68	85	62	68
62	58	94	91
66	79	79	56
62	57	86	89
60	88	85	94

SS(treatment): _____   MS(treatment): _____   Test statistic F: _____

SS(error): _____   MS(error): _____ P-value: _____

SS(total): _____

Conclusion: _____

_____

11-5. **Mean Weights of Different Colas** Refer to Data Set 1 in Appendix B. Is there sufficient evidence to support the claim that cans of regular Coke, diet Coke, regular Pepsi, and diet Pepsi have different mean weights? (The four samples are already STATDISK files that can be opened for use. The file names are CKREGWT (Coke regular weight), CKDIETWT (Coke diet weight), PPREGWT (Pepsi regular weight), and PPDIETWT (Pepsi diet weight).

SS(treatment): _____   MS(treatment): _____   Test statistic F: _____

SS(error): _____   MS(error): _____ P-value: _____

SS(total): _____

Conclusion: _____

_____

If there is a difference, can you suggest a feasible explanation?

_____

_____

11-6.   **Mean Weights of M&Ms**   Refer to Data Set 10 in Appendix B. At the 0.05 significance level, test the claim that the mean weight of M&Ms is the same for each of the six different color populations. The STATDISK files are RED, ORANGE, YELLOW, BROWN, BLUE, and GREEN.

SS(treatment): _____   MS(treatment): _____   Test statistic F: _____

SS(error):   _____   MS(error):   _____ P-value: _____

SS(total):   _____

Conclusion:_____

_____

If it is the intent of Mars, Inc. to make the candies so that the different color populations have the same mean, do these results suggest that the company has a problem requiring corrective action?

_____

11-7.   **Second-Hand Smoke in Different Groups**   Refer to Data Set 9 in Appendix B. Use a 0.01 significance level to test the claim that the mean cotinine level is different for these three groups: nonsmokers not exposed to environmental tobacco smoke (NETS), nonsmokers who are exposed to tobacco smoke (ETS), and people who smoke (SMOKERS).

SS(treatment): _____   MS(treatment): _____   Test statistic F: _____

SS(error):   _____   MS(error):   _____ P-value: _____

SS(total):   _____

Conclusion:_____

_____

What do the results suggest about "second-hand" smoke?

_____

11-8.   **Flammability Testing**   Flammability tests were conducted on children's sleepwear. The Vertical Semi-restrained Test was used by burning a piece of fabric under very controlled conditions. After the burning stops, the length of the charred portion is measured and recorded. Results are given below for the same fabric tested at different laboratories. Because the same fabric is used, the different laboratories should provide the same results. Do they? Use a 0.05 significance level to test the claim that the different laboratories have the same population mean. (The data were provided by Minitab.)

Laboratory

1	2	3	4	5
2.9	2.7	3.3	3.3	4.1
3.1	3.4	3.3	3.2	4.1
3.1	3.6	3.5	3.4	3.7
3.7	3.2	3.5	2.7	4.2
3.1	4.0	2.8	2.7	3.1
4.2	4.1	2.8	3.3	3.5
3.7	3.8	3.2	2.9	2.8
3.9	3.8	2.8	3.2	
3.1	4.3	3.8	2.9	
3.0	3.4	3.5		
2.9	3.3			

SS(treatment): _____   MS(treatment): _____   Test statistic F: _____

SS(error): _____   MS(error): _____ P-value: _____

SS(total): _____

Conclusion: _____

_____

11-9. **Leader Longevity** See the Technology Project at the end of Chapter 11 in *Elementary Statistics*, where data are listed for presidents, popes and British monarchs. The data are the numbers of years that they lived after their inauguration, election, or coronation (based on data from *Computer-Interactive Data Analysis* by Lunn and McNeil, John Wiley & Sons). Use boxplots and analysis of variance to determine whether the survival times for the different groups differ. Obtain printed copies of the computer displays and write your observations and conclusions.

_____

_____

_____

11-10. **Simulations** Use STATDISK to randomly generate three different samples of 500 values each.(Select **Data,** then **Normal Generator**.) For the first two samples, use a normal distribution with a mean of 100 and a standard deviation of 15. For the third sample, use a normal distribution with a mean of 101 and a standard deviation of 15. We know that the three populations have different means, but do the methods of analysis of variance allow you to conclude that the means are different? Explain.

_____

_____

_____

12

Statistical Process Control

The major topics of Chapter 12 from *Elementary Statistics* are run charts, control charts for variation, control charts for mean, and control charts for attributes. Although STATDISK is not programmed to generate run charts or control charts, there are ways to obtain them by using STATDISK's scatterplot feature.

12-1 Run Charts

In Chapter 12 from *Elementary Statistics* we define **process data** to be data arranged according to some time sequence, such as the data in Table 12-1, which is reproduced below. In Table 12-1, each workday is shown with seven measured axial loads of aluminum cans. An axial load is the maximum weight the sides can support. The aluminum cola cans are manufactured without the tops, they are filled, then the tops are pressed into place. It is therefore important that the cans be strong enough (with a sufficiently large axial load) to support the pressure applied when the top lids are pressed into place with pressures that vary between 158 pounds and 165 pounds.

Table 12-1 Axial loads (in pounds) of Aluminum Cans

Day	Axial Load (pounds)							Mean	Median	Range	St. Dev.
1	270	273	258	204	254	228	282	252.7	258	78	27.6
2	278	201	264	265	223	274	230	247.9	264	77	29.7
3	250	275	281	271	263	277	275	270.3	275	31	10.6
4	278	260	262	273	274	286	236	267.0	273	50	16.3
5	290	286	278	283	262	277	295	281.6	283	33	10.7
6	274	272	265	275	263	251	289	269.9	272	38	11.8
7	242	284	241	276	200	278	283	257.7	276	84	31.4
8	269	282	267	282	272	277	261	272.9	272	21	7.9
9	257	278	295	270	268	286	262	273.7	270	38	13.5
10	272	268	283	256	206	277	252	259.1	268	77	25.9
11	265	263	281	268	280	289	283	275.6	280	26	10.1
12	263	273	209	259	287	269	277	262.4	269	78	25.3
13	234	282	276	272	257	267	204	256.0	267	78	27.8
14	270	285	273	269	284	276	286	277.6	276	17	7.3
15	273	289	263	270	279	206	270	264.3	270	83	27.0
16	270	268	218	251	252	284	278	260.1	268	66	22.3
17	277	208	271	208	280	269	270	254.7	270	72	32.2
18	294	292	289	290	215	284	283	278.1	289	79	28.1
19	279	275	223	220	281	268	272	259.7	272	61	26.5
20	268	279	217	259	291	291	281	269.4	279	74	25.9
21	230	276	225	282	276	289	288	266.6	276	64	27.2
22	268	242	283	277	285	293	248	270.9	277	51	19.3
23	278	285	292	282	287	277	266	281.0	282	26	8.4
24	268	273	270	256	297	280	256	271.4	270	41	14.3
25	262	268	262	293	290	274	292	277.3	274	31	14.1

To use STATDISK for generating a run chart, pair the data with the positive integers, then generate a scatter diagram (available in the Correlation and Regression module). The table below shows the first row of data values paired with 1, 2, 3, ..., 7.

x	1	2	3	4	5	6	7
y	270	273	258	204	254	228	282

This table includes only the first seven values in Table 12-1, but it can be easily extended to include more data. To include the entries from day 2, for example, continue with the positive integers and enter the row 2 data as shown below.

x	8	9	10	11	12	13	14
y	278	201	264	265	223	274	230

Continue with this procedure to enter the consecutive table values matched with the positive integers. The paired data can be entered in the Sample Editor module, or they can be entered directly in the Correlation and Regression module. (See Section 9-1 of this manual/workbook.)

Shown below is the scatter diagram for the first 50 entries of Table 12-1 paired with the positive integers 1, 2, 3, ..., 50.

STATDISK Run Chart

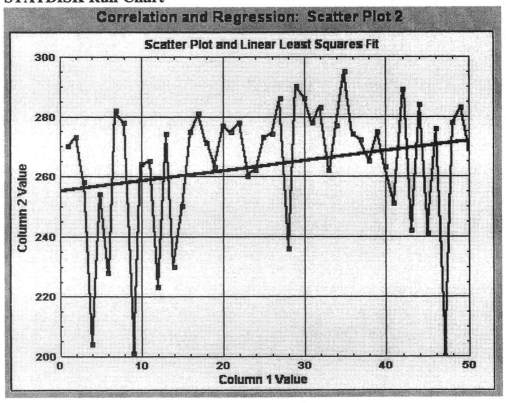

STATDISK does not connect the points as shown in the above display, but they can be easily connected as shown. If you extend the table to include all 175 values, and ignore the regression line, you will effectively create the same run chart as Figure 12-1 in Section 12-2 of the textbook.

STATDISK Procedure for Run Chart

Based on the preceding example, here are the steps for using STATDISK to create a run chart:

1. Select **Analysis** from the main menu.

2. Select the menu item of **Correlation and Regression** (to get a scatter diagram).

3. Proceed to enter the positive consecutive integers 1, 2, 3, ... in column 1. Also, enter the consecutive sample data in column 2.

4. Click on the **Evaluate** bar, but ignore the results.

5. Click on **Plot 1** to get the scatter diagram.

6. Connect the points in order from left to right, as shown in the above run chart. Ignore the regression line included with the graph.

To *interpret* the run chart, use the criteria described in Section 12-2 of the textbook. See Figure 12-2 for typical samples of process patterns that are *not* statistically stable.

12-2   Control Charts for Variation

Section 12-2 of the textbook describes *R* **charts** (sequential plots of ranges) for monitoring *variation* in a process. Using the data of Table 12-1, for example, the *R* chart is a plot of the ranges 78, 77, ..., 31. Again, STATDISK does not have a feature designed specifically for *R* charts, but they can be obtained as follows.

STATDISK Procedure for *R* Chart to Monitor Process Variation

1. Find the value of the range for each individual sample.

2. Determine the values to be used for the upper control limit, the centerline, and the lower control limit. (See Section 12-2 in *Elementary Statistics*.)

3. Select **Analysis** from the main menu.

4. Select the menu item of **Correlation and Regression** (to get a scatter diagram).

5. Proceed to enter these values in the first column: 0, 0, 0, 1, 2, 3, ...

6. In the second column, enter the value of the upper control limit, then the value for the centerline, then the value for the lower control limit, followed by the values of the sample ranges. That is, follow the general format given on the top of the next page.

x	y
0	(Value of upper control limit)
0	(Value $\bar{R}$ for the centerline)
0	(Value of lower control limit)
1	(Value of first sample range)
2	(Value of second sample range)
3	(Value of third sample range)
and so on	

Using the sample ranges in Table 12-1 along with the locations of the upper control limit (105.74), the centerline (54.96), and the lower control limit (4.18), the STATDISK dialog box should appear as shown below. (For the procedure used to find these values of 105.74, 54.96, and 4.18, see Section 12-2 of *Elementary Statistics*.)

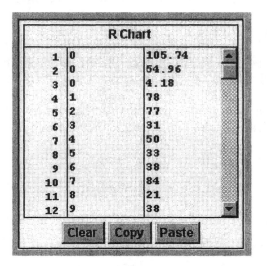

7.   Click on the **Evaluate** bar, but ignore the results.

8.   Click on **Plot 1** to get the scatter diagram.

9.   The three leftmost points will be stacked above 0. Use those points to position the upper control limit, the centerline, and the lower control limit. For the remaining points beginning above 1, connect the points in order from left to right, and ignore the regression line included with the graph.

Shown on the next page is the STATDISK R chart for the data in Table 12-1. The graph was modified to include lines and labels for the upper control limit, the centerline, and the lower control limit. Note that the three thick lines were located by including these three points as the first three points in the list of paired data: (0, 105.74), (0, 54.96), (0, 4.18).

We can interpret the R chart by applying the three out-of-control criteria given in the textbook. We conclude that variation in this process is within statistical control because of the following.

1. There is no pattern, trend, or cycle that is obviously not random.

2. No point lies beyond the upper or lower control limits.

3. There are not 8 consecutive points all above or all below the center line.

STATDISK *R* Chart

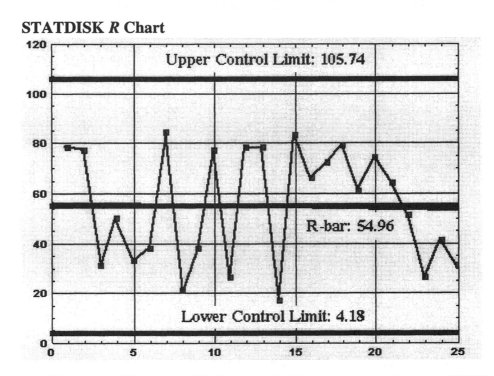

12-3 Control Charts for Mean

Control charts for $\bar{x}$ can be constructed by using the same methods described for *R* charts. You must first determine the sample means and the values to be used for the upper control limit, the centerline, and the lower control limit. See Section 12-2 in *Elementary Statistics* for the procedures that can be used to find these values.

STATDISK Procedure for $\bar{x}$ Chart to Monitor Process Mean

1.   Find the value of the mean for each individual sample.

2.   Determine the values to be used for the upper control limit, the centerline, and the lower control limit. (See Section 12-2 in *Elementary Statistics*.)

3.   Select **Analysis** from the main menu.

4.   Select the menu item of **Correlation and Regression** (to get a scatter diagram).

5.   Proceed to enter these values in the first column: 0, 0, 0, 1, 2, 3, ...

6.   In the second column, enter the value of the upper control limit, then the value for the centerline, then the value for the lower control limit, followed by the

values of the sample means. Using the sample means in Table 12-1 along with the locations of the upper control limit (290.15), the centerline (267.12), and the lower control limit (244.09), the STATDISK dialog box should appear as shown below. (For the procedure used to find these values of 290.15, 267.12, and 244.09, see Section 12-2 of *Elementary Statistics*.)

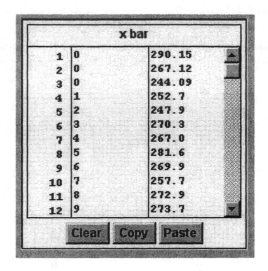

7.   Click on the **Evaluate** bar, but ignore the results.

8.   Click on **Plot 1** to get the scatter diagram.

9.   The three leftmost points will be stacked above 0. Use those points to position the upper control limit, the centerline, and the lower control limit. For the remaining points beginning above 1, connect the points in order from left to right, and ignore the regression line included with the graph.

Shown on the next page is the STATDISK $\bar{x}$ chart for the data in Table 12-1. The graph was modified to include lines and labels for the upper control limit, the centerline, and the lower control limit. Note that the three thick lines were located by including these three points as the first three points in the list of paired data: (0, 290.15), (0, 267.12), (0, 244.09).

We can interpret the $\bar{x}$ chart by applying the three out-of-control criteria given in the textbook. We conclude that the mean in this process is within statistical control because of the following.

1. There is no pattern, trend, or cycle that is obviously not random.

2. No point lies beyond the upper or lower control limits.

3. There are not 8 consecutive points all above or all below the center line.

STATDISK $\bar{x}$ Chart

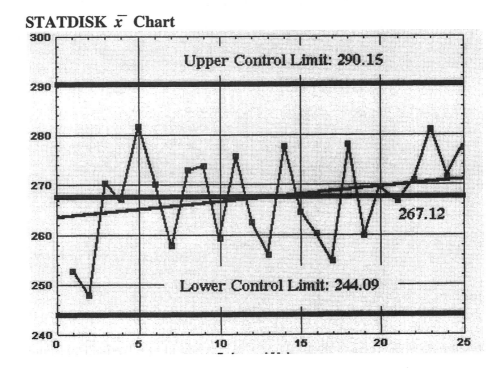

12-4 Control Charts for Attributes

A control chart for attributes (or p chart) can also be constructed by using the same procedure for R charts and $\bar{x}$ charts. A p chart is very useful in monitoring some process proportion, such as the proportions of defects over time. The example in Section 12-3 of the textbook involves the numbers of deaths from respiratory tract infections among 100,000 people in each of 13 consecutive years, with the results listed below (based on data from "Trends in Infectious Diseases Mortality in the United States" by Pinner and others, *Journal of the American Medical Association*, Vol. 275, No. 3).

Number of deaths: 25 24 22 25 27 30 31 30 33 32 33 32 31

Refer to the example in Section 12-3 of the textbook and use the following procedure:

STATDISK Procedure for p Charts

1. Collect the list of sample proportions. For the above sample data, the proportions are 0.00025, 0.00024, ..., 0.00031.

2. Determine the values to be used for the upper control limit, the centerline, and the lower control limit. (See Section 12-3 in *Elementary Statistics*.) For the above sample data, the upper control limit is at 0.000449, the centerline is at 0.000288, and the lower control limit is at 0.000127.

3. Select **Analysis** from the main menu.

4. Select the menu item of **Correlation and Regression** (to get a scatter diagram).

5.   Proceed to enter these values in the first column: 0, 0, 0, 1, 2, 3, ...

6.   In the second column, enter the value of the upper control limit, then the value for the centerline, then the value for the lower control limit, followed by the values of the sample proportions. For the example in Section 12-3 of the textbook we should make entries in the dialog box as shown below.

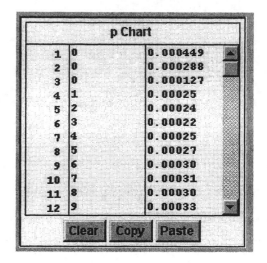

7.   Click on the **Evaluate** bar, but ignore the results.

8.   Click on **Plot 1** to get the scatter diagram.

9.   The three leftmost points will be stacked above 0. Use those points to position the upper control limit, the centerline, and the lower control limit. For the remaining points beginning above 1, connect the points in order from left to right, and ignore the regression line included with the graph.

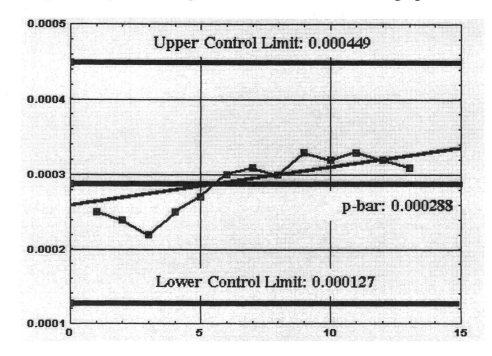

The above control chart for p can be interpreted by using the same three out-of-control criteria listed in Section 12-2 of the textbook. Using those criteria, we concluded that this process is out of statistical control because there appears to be an upward trend, and there are eight consecutive points all lying above the center line.

Minitab and some other statistical software packages do have the ability to directly generate run charts and control charts that are so important for monitoring process data over time. Shown below is the Minitab p chart that is automatically generated with lines for the upper control limit, centerline, and lower control limit. The use of such charts is increasing as more businesses recognize that this statistical tool can be effective in increasing quality and lowering costs.

Minitab p Chart

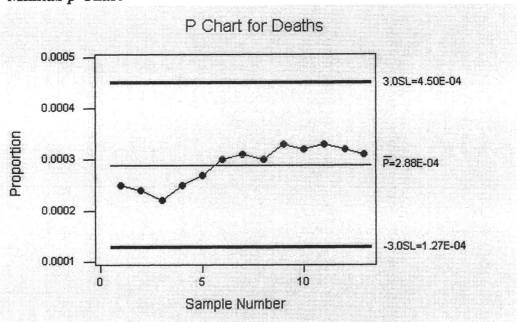

CHAPTER 12 EXPERIMENTS: Statistical Process Control

12-1. **Old Faithful Geyser** Exercise 1 in Section 12-2 from *Elementary Statistics* includes process data from the Old Faithful geyser, which was monitored for 25 recent and consecutive years. In each year, six intervals (in minutes) between eruptions were recorded. Use STATDISK to generate a run chart of the individual values. Does there appear to be a pattern suggesting that the process is not within statistical control?

Year	Interval (min)						Mean	Range
1	65	72	60	69	65	67	66.3	12
2	74	65	60	69	68	59	65.8	15
3	68	66	69	64	70	73	68.3	9
4	73	65	71	77	63	77	71.0	14
5	79	67	64	61	81	77	71.5	20
6	74	76	65	69	76	64	70.7	12
7	70	73	74	77	65	73	72.0	12
8	71	68	70	79	75	82	74.2	14
9	62	63	61	48	59	77	61.7	29
10	60	74	77	57	52	78	66.3	26
11	67	73	47	81	92	57	69.5	45
12	79	84	79	72	61	80	75.8	23
13	83	78	83	74	61	68	74.5	22
14	57	68	72	75	56	79	67.8	23
15	59	76	78	86	64	72	72.5	27
16	63	63	71	77	81	65	70.0	18
17	67	84	72	75	70	70	73.0	17
18	93	83	85	79	90	74	84.0	19
19	81	74	80	65	70	84	75.7	19
20	83	67	71	67	97	88	78.8	30
21	62	61	57	86	70	77	68.8	29
22	67	75	67	89	93	81	78.7	26
23	86	65	70	74	83	74	75.3	21
24	74	67	99	75	41	83	73.2	58
25	97	93	73	81	85	90	86.5	24

12-2.   **Old Faithful Geyser: Constructing an _R_ Chart**   Using the same process data from Experiment 12-1, construct an _R_ chart and determine whether the process variation is within statistical control. If it is not, identify which of the three out-of-control criteria lead to rejection of statistically stable variation.

12-3.   **Old Faithful Geyser: Constructing an $\bar{x}$ Chart**   Using the same process data from Experiment 12-1, construct an $\bar{x}$ chart and determine whether the process mean is within statistical control. If it is not, identify which of the three out-of-control criteria lead to rejection of a statistically stable mean.

12-4.   **Weights of Minted Quarters**   Exercise 4 in Section 12-2 from _Elementary Statistics_ includes the weights (in grams) of newly minted quarters. The U.S. Mint has a goal of making quarters with a weight of 5.670 g, but any weight between 5.443 g and 5.897 g is considered acceptable. A new minting machine is placed into service and the weights are recorded for a quarter randomly selected every 12 min for 20 consecutive hours. The results are listed in the accompanying table. Use STATDISK to construct a run chart. Determine whether the process appears to be within statistical control.

Hour	Weight (grams)					Mean	Range
1	5.639	5.636	5.679	5.637	5.691	5.6564	0.055
2	5.655	5.641	5.626	5.668	5.679	5.6538	0.053
3	5.682	5.704	5.725	5.661	5.721	5.6986	0.064
4	5.675	5.648	5.622	5.669	5.585	5.6398	0.090
5	5.690	5.636	5.715	5.694	5.709	5.6888	0.079
6	5.641	5.571	5.600	5.665	5.676	5.6306	0.105
7	5.503	5.601	5.706	5.624	5.620	5.6108	0.203
8	5.669	5.589	5.606	5.685	5.556	5.6210	0.129
9	5.668	5.749	5.762	5.778	5.672	5.7258	0.110
10	5.693	5.690	5.666	5.563	5.668	5.6560	0.130
11	5.449	5.464	5.732	5.619	5.673	5.5874	0.283
12	5.763	5.704	5.656	5.778	5.703	5.7208	0.122
13	5.679	5.810	5.608	5.635	5.577	5.6618	0.233
14	5.389	5.916	5.985	5.580	5.935	5.7610	0.596
15	5.747	6.188	5.615	5.622	5.510	5.7364	0.678
16	5.768	5.153	5.528	5.700	6.131	5.6560	0.978
17	5.688	5.481	6.058	5.940	5.059	5.6452	0.999
18	6.065	6.282	6.097	5.948	5.624	6.0032	0.658
19	5.463	5.876	5.905	5.801	5.847	5.7784	0.442
20	5.682	5.475	6.144	6.260	6.760	6.0642	1.285

12-5. **Minting Quarters: Constructing an *R* Chart**   Using the same process data from Experiment 12-4, construct an *R* chart and determine whether the process variation is within statistical control. If it is not, identify which of the three out-of-control criteria lead to rejection of statistically stable variation.

12-6. **Minting Quarters: Constructing an $\bar{x}$ Chart**   Using the same process data from Experiment 12-4, construct an $\bar{x}$ chart and determine whether the process mean is within statistical control. If it is not, identify which of the three out-of-control criteria lead to rejection of a statistically stable mean. Does this process need corrective action?

12-7. **Home Energy Consumption: Run Chart**   Exercise 7 in Section 12-2 from *Elementary Statistics* includes the electrical energy consumption (in kilowatt-hours) amounts for the author's home in upstate New York. Process data are listed for two-month intervals over four years. Construct a run chart for the 24 values. Does there appear to be a pattern suggesting that the process is not within statistical control? Is there any pattern or variation that can be explained?

Year	Jan.-Feb.	March-April	May-June	July-Aug.	Sept.-Oct.	Nov.-Dec.
1	4762	3875	2657	4358	2201	3187
2	4504	3237	2198	2511	3020	2857
3	3952	2785	2118	2658	2139	3071
4	3863	3013	2023	2953	3456	2647

12-8. **Home Energy Consumption: *R* Chart**   Refer to the same process data from Experiment 12-7. Use samples of size three by combining the first three values for each year and combining the last three values for each year. Construct an *R* chart and determine whether the process variation is within statistical control. If it is not, identify which of the three out-of-control criteria lead to rejection of statistically stable variation.

12-9. **Home Energy Consumption: Constructing an $\bar{x}$ Chart**   Refer to the same process data from Experiment 12-7. Use samples of size three by combining the first three values for each year and combining the last three values for each year. Construct an $\bar{x}$ chart and determine whether the process mean is within statistical control. If it is not, identify which of the three out-of-control criteria lead to rejection of a statistically stable mean. What is a practical effect of not having this process in statistical control? Give an example of a cause that would make the process go out of statistical control.

12-10. ***p* Chart for Deaths from Infectious Diseases**   In Exercise 5 from Section 12-3 of *Elementary Statistics,* it is stated that in each of 13 consecutive and recent years 100,000 children aged 0–4 years were randomly selected and the number who died from infectious diseases is recorded, with the results given below (based on data from "Trends in Infectious Diseases Mortality in the United States," by Pinner et al., *Journal of the American Medical Association,* Vol. 275, No. 3). Use STATDISK to construct a *p* chart. Do the results suggest a problem that should be corrected?

Number who died: 30 29 29 27 23 25 25 23 24 25 25 24 23

12-11. **_p_ Chart for Victims of Crime** Exercise 6 in Section 12-3 from _Elementary Statistics_ states that for each of 20 consecutive and recent years, 1000 adults were randomly selected and surveyed. Each value below is the number who were victims of violent crime (based on data from the U.S. Department of Justice, Bureau of Justice Statistics). Use STATDISK to construct a _p_ chart. Do the data suggest a problem that should be corrected?

29 33 24 29 27 33 36 22 25 24 31 31 27 23 30 35 26 31 32 24

12-12. **_p_ Chart for Boston Rainfall** Exercise 7 in Section 12-3 of _Elementary Statistics_ refers to the Boston rainfall amounts in Data Set 17 of Appendix B. Delete the last value for Wednesday, so that there are 52 weeks of seven days each. For each of the 52 weeks, let the sample proportion be the proportion of days that it rained. In the first week for example, the sample proportion is $3/7 = 0.429$. Use STATDISK to construct a _p_ chart. Do the data represent a statistically stable process?

12-13. **_p_ Chart for Marriage Rates** Do Exercise 8 in Section 12-3 of _Elementary Statistics_. That exercise asks that _p_ charts be used to compare the statistical stability of the marriage rates of Japan and the United States. In each year, 10,000 people in each country were randomly selected, and the numbers of marriages are given for eight consecutive and recent years (based on United Nations data).

| Japan: | 58 60 61 64 63 63 64 63 |
| United States: | 98 94 92 90 91 89 88 87 |

12-14. **Effects of Transformations** It is often easier to handle data by transforming them in some way. Given the data in Experiment 12-4, for example, it would be easier to omit the decimal points, which is equivalent to multiplying each weight by 1000. If each value is multiplied by 1000, how are the run chart, _R_ chart, and $\bar{x}$ chart affected?

Run chart: _____

_R_ chart: _____

$\bar{x}$ chart: _____

12-15. **Effect of Transformation** Experiment 12-14 considered a transformation consisting of multiplying every value by the same constant. How are the run chart, _r_ chart, and $\bar{x}$ chart affected if the same constant is added to (or subtracted from) every value?

Run chart: _____

_R_ chart: _____

$\bar{x}$ chart: _____

13

Nonparametric Statistics

13-1 Nonparametric Methods

STATDISK includes a wide variety of nonparametric procedures. It includes procedures for all of the nonparametric methods described in Chapter 13 of *Elementary Statistics*. The sections of this chapter correspond to those in the textbook.

Section 13-1 in the textbook introduces some basic principles of nonparametric methods. The textbook notes that prior to Chapter 13 most of the methods of inferential statistics are called *parametric* methods, because they are based on sampling from a population with specific parameters (such as the mean μ, standard deviation σ, or proportion p). Those parametric methods usually have some fairly strict conditions, such as a requirement that the sample data must come from a normally distributed population. Because nonparametric methods do not require specific distributions (such as the normal distribution), these nonparametric methods are often called **distribution-free tests**. The following sections describe some of the more important and commonly-used nonparametric methods.

13-2 Sign Test

Section 13-2 in *Elementary Statistics* includes the following definition.

Definition
The **sign test** is a nonparametric (distribution-free) test that uses plus and minus signs to test different claims, including these:

1. Given matched pairs of sample data, test claims about the medians of the two populations.
2. Given nominal data, test claims about the proportion of some category.
3. Test claims about the median of a single population.

STATDISK makes it possible to work with all three of the above cases. We will first describe the STATDISK procedure, then we will illustrate it.

STATDISK Procedure for the Sign Test

1. Select **Analysis** from the main menu.

2. Select **Sign Tests** from the submenu.

3. If you already know the number of positive and negative signs (such as cases involving nominal data), select the option of **Given Number of Signs**.

   If you have sample *paired* data, select the option of **Given Pairs of Values**.

4. The content of the dialog box will depend on the choice made in step 3. Both cases require that you select the form of the claim being tested and a significance level such as 0.05 or 0.01. You must then enter the numbers of positive and negative signs, or enter the original pairs of data.

5. Click on the **Evaluate** bar.

6. Click on **Plot** to obtain a graph that includes the test statistic and critical value. The plot will be generated only if the normal approximation is used (because $n > 25$).

Section 13-2 includes the data in Table 13-2, which is reproduced below. The values represent the reported and measured heights of 12 males statistics students, and we want to test the claim that there is no difference. The STATDISK results are shown below, which shows that the test statistic is $x = 1$, the critical value (found from Table A-7) is $x = 2$, the null hypothesis is rejected. There is sufficient sample evidence to warrant rejection of the claim that there is no difference between reported and measured heights. There does appear to be a difference.

Table 13-2   Reported and Measured Heights of Male Statistics Students

Reported Height	68	74	82.25	66.5	69	68	71	70	70	67	68	70
Measured Height	66.8	73.9	74.3	66.1	67.2	67.9	69.4	69.9	68.6	67.9	67.6	68.8
Difference	1.2	0.1	7.95	0.4	1.8	0.1	1.6	0.1	1.4	-0.9	0.4	1.2
Sign of difference	+	+	+	+	+	+	+	+	+	-	+	+

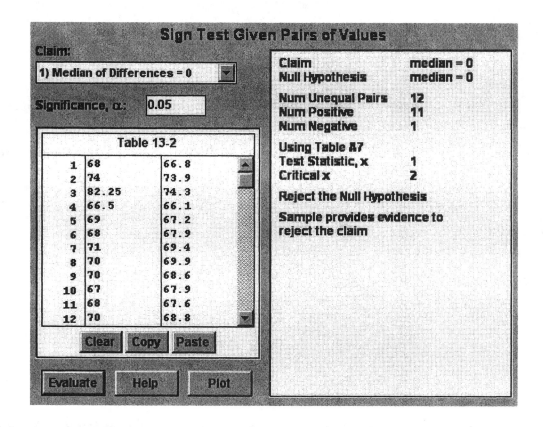

The preceding example uses the original paired sample data, but STATDISK does allow you to enter the number of positive signs and the number of negative signs if those numbers are know. The textbook includes an example involving the hiring of 30 men and 70 women. Using STATDISK, we could select the option of **Given Number of Signs**. (In fact, you might often find it easier to manually determine the signs than to enter the original paired data.)

13-3 Wilcoxon Signed-Ranks Test

Section 13-3 in *Elementary Statistics* describes the Wilcoxon signed-ranks test, and the following definition is given.

Definition
The **Wilcoxon signed-ranks test** is a nonparametric test that uses ranks of sample data consisting of *matched pairs*. It is used to test for differences in the population distributions, so the null and alternative hypotheses are as follows:
H_0: The two samples come from populations with the same distribution.
H_1: The two samples come from populations with different distributions.

The textbook makes this important point: The Wilcoxon signed-ranks test and the sign test can both be used with sample data consisting of matched pairs, but the sign test uses only the signs of the differences and not their actual magnitudes (how large the numbers are). The Wilcoxon signed-ranks test uses ranks, so the magnitudes of the differences are taken into account. Because the Wilcoxon signed-ranks test incorporates and uses more information than the sign test, it tends to yield conclusions that better reflect the true nature of the data. First we describe the STATDISK procedure for conducting a Wilcoxon signed-ranks test, then we illustrate it with an example.

STATDISK Procedure for the Wilcoxon Signed-Ranks Test

1. Select **Analysis** from the main menu.

2. Select **Wilcoxon Tests** from the submenu.

3. You must now choose between the following two options.

   Signed-Ranks Test (Select this option with *matched pairs*.)
   Rank-Sum Test   (Select this option with two *independent* samples)

4. After selecting the signed-ranks test option, proceed to enter a significance level, such as 0.05 or 0.01.

5. Enter the matched sample data in the two columns.

6. Click on the **Evaluate** bar.

7. Click on the **Plot** bar to see a graph that includes the test statistic and critical values. The graph will be displayed only if the normal approximation is used (because $n > 30$).

Section 13-3 of the textbook includes an example that uses the same sample data listed in Table 13-2, shown in the preceding example of this manual/workbook. The next page shows the STATDISK display for this sample. Based on that display, we see that the test statistic is $T = 6$, the critical value is $T = 14$, and the conclusion is to reject the null hypothesis. There is sufficient evidence to warrant rejection of the null hypothesis of no difference between the population of reported heights and the population of measured heights. As with the sign test, we conclude that there is a difference between reported heights and measured heights.

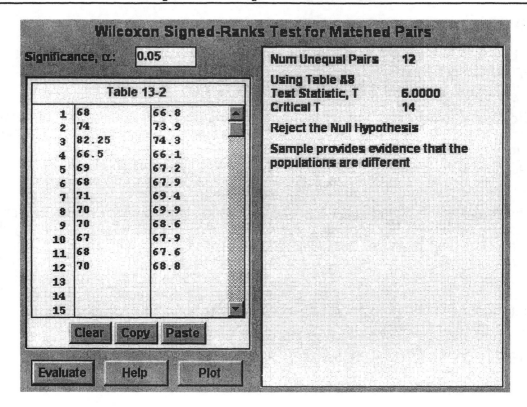

13-4 Wilcoxon Rank-Sum Test

Section 13-4 of *Elementary Statistics* includes the following definition.

Definition
The **Wilcoxon rank-sum test** is a nonparametric test that uses ranks of sample data from two *independent* populations. It is used to test the null hypothesis that the two independent samples come from populations with the same distribution. (That is, the two populations are identical.) The alternative hypothesis is the claim that the two population distributions are different in some way.

H_0:   The two samples come from populations with the same distribution. (That is, the two populations are identical.)

H_1:   The two samples come from populations with different distributions. (That is, the two populations are different in some way.)

First we describe the STATDISK procedure for conducting a Wilcoxon rank-sum test, then we illustrate it with an example.

STATDISK Procedure for the Wilcoxon Rank-Sum Test

1.   Select **Analysis** from the main menu.

2.   Select **Wilcoxon Tests** from the subdirectory.

3.   You must now choose between the following two options.

Signed-Ranks Test (Select this option with *matched pairs*.)
Rank-Sum Test (Select this option with two *independent* samples.)

4. After selecting the rank-sum test option, proceed to enter a significance level, such as 0.05 or 0.01.

5. Enter the values of the first sample in the first column, then enter the values of the second sample in the second column. (If the data sets are large, it might be wise to use the Sample Editor to enter and save the data sets; they can then be copied to the Wilcoxon rank-sum test module by using Copy/Paste.)

6. Click on the **Evaluate** bar.

7. Click on **Plot** to display a graph that shows the test statistic and critical values.

Section 13-4 of the textbook includes an example illustrating the Wilcoxon rank-sum test applied to the Boston rainfall amounts for Wednesday and Saturday. Those two data sets are listed in Data Set 17 of Appendix B from the textbook; they are also available in STATDISK as the data files named rainwed.sdd and rainsat.sdd. Instead of manually entering those data sets, simply retrieve them in STATDISK by using File/Open and Copy/Paste. The STATDISK results are shown below. The rank sum for Wednesday is 2639 and the rank sum for Saturday is 2926. The test statistic is $z = -1.0895$ and the critical values are $z = -1.96$ and $z = 1.96$. We fail to reject the null hypothesis of no difference. We conclude that the differences between Wednesday and Saturday are not significant.

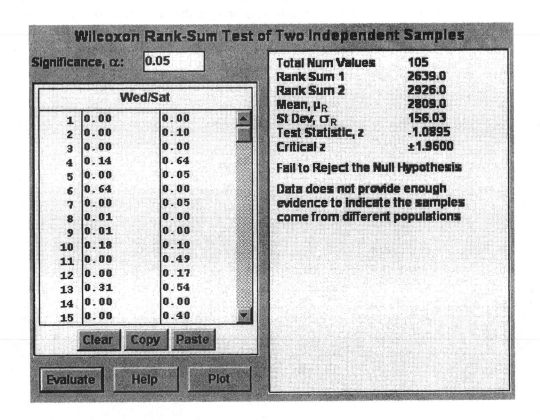

13-5 Kruskal-Wallis Test

Section 13-5 of *Elementary Statistics* includes this definition.

Definition
The **Kruskal-Wallis test** (also called the *H* test) is a nonparametric test that uses ranks of sample data from three or more independent populations. It is used to test the null hypothesis that the independent samples come from populations with the same distribution; the alternative hypothesis is the claim that the two population distributions are different in some way.

H_0: The samples come from populations with the same distribution.
H_1: The samples come from populations with different distributions.

We describe the STATDISK procedure for the Kruskal-Wallis test, then we illustrate it with an example.

STATDISK Procedure for the Kruskal-Wallis Test

1.   Select **Analysis** from the main menu.

2.   Select **Kruskal-Wallis test** from the submenu.

3.   In the dialog box, enter a significance level, such as 0.05 or 0.01.

4.   Enter the values of the first sample in the first column, then enter the values of the second sample in the second column, and so on. (If the data sets are large, it might be wise to use the Sample Editor to enter and save the data sets; they can then be copied to the Kruskal-Wallis test module by using Copy/Paste.) After entering the first three columns, use the > key to move to the right for the fourth column and any additional columns.

5.   Click on the **Evaluate** bar.

6.   Click on **Plot** to display a graph that shows the test statistic and critical values.

The textbook illustrates the Kruskal-Wallis test with the sample data in Table 11-1, which is included in Section 11-1 of this manual/workbook. We use STATDISK's Kruskal-Wallis test to determine whether there is sufficient evidence to conclude that the *head injuries* for the four car weight categories are the same. After entering the sample data, the results are as shown in the display on the next page. The display shows only the first three columns of sample data, but the fourth column is also included. Use the > key and the < key to move to the right and left for columns that are not visible.

Important elements of the STATDISK display include the rank sums of 64, 52.5, 44.5, 49, the test statistic of $H = 1.1914$, and the critical value of $H = 7.8147$. Remember that the Kruskal-Wallis test is a *right-tailed* test, so we reject the null hypothesis (the samples come from populations with the same distribution) only if the value of the test statistic is equal to or greater than the critical value. Because the test statistic of $H = 1.1914$ is less than $H = 7.8147$, we fail to reject the null hypothesis. There is not sufficient sample evidence to support the claim that the head injury measurements are different for the various car weight categories.

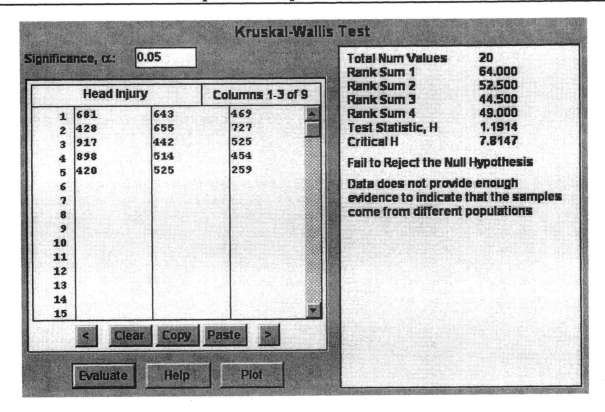

13-6 Rank Correlation

Section 13-6 of *Elementary Statistics* introduces the use of *rank correlation*, which uses ranks in a procedure for determining whether there is some relationship between two variables.

Definition
The **rank correlation test** (or Spearman's rank correlation test) is a nonparametric test that uses ranks of sample data consisting of matched pairs. It is used to test for an association between two variables, so the null and alternative hypotheses are as follows (where ρ_s denotes the rank correlation coefficient for the entire population):

H_0: $\rho_s = 0$ (There is *no* correlation between the two variables.)
H_1: $\rho_s \neq 0$ (There is a correlation between the two variables.)

First we describe the STATDISK procedure, then we illustrate it with an example.

STATDISK Procedure for Rank Correlation

1.   Select **Analysis** from the main menu.

2.   Select **Rank Correlation** from the submenu.

3.   Enter a significance level, such as 0.05 or 0.01.

4.   Enter the paired data in columns 1 and 2. (If there are many pairs of data, consider using the Sample Editor module to enter and save the data, which can then be copied to the rank correlation module by using Copy/Paste.)

5.   Click on the **Evaluate** bar.

6.   Click on the **Plot** bar to obtain a graph that shows the test statistic and critical values. The graph will be displayed only if the normal approximation is used (because $n > 30$).

Section 13-6 of the textbook includes paired data describing the *Business Week* rankings of business schools that were ranked by corporations and by MBA graduates. The sample data are listed in Table 13-6, and the STATDISK results follow. The STATDISK display includes the rank correlation coefficient of $r_s = 0.10303$ and the critical values of $r_s = 0.648$ and $r_s = -0.648$. We fail to reject the null hypothesis of $\rho_s = 0$ (*no* correlation). There is not sufficient evidence to conclude that there is a difference between the corporate rankings and graduate rankings of business schools.

Table 13-6 Rankings of Business Schools

	PA	NW	Chi	Sfd	Hvd	MI	IN	Clb	UCLA	MIT
Corporate rank	1	2	4	5	3	6	8	7	10	9
Graduate rank	3	5	4	1	10	7	6	8	2	9

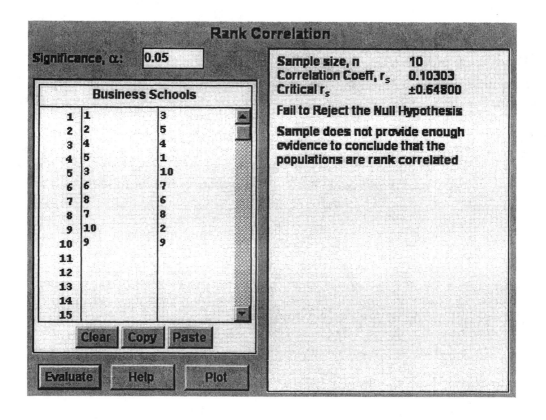

13-7 Runs Test for Randomness

Section 13-7 of *Elementary Statistics* includes these definitions.

Definitions
A **run** is a sequence of data having the same characteristic; the sequence is preceded and followed by data with a different characteristic or by no data at all.
The **runs test** uses the number of runs in a sequence of sample data to test for randomness in the order of the data.

First we describe the STATDISK procedure for the runs test for randomness, then we illustrate it with an example.

STATDISK Procedure for the Runs Test for Randomness

1. Using the original data, count the number of runs, the number of elements of the first type, and the number of elements of the second type.

2. Select **Analysis** from the main menu.

3. Select **Runs Test** from the submenu.

4. Make these entries in the dialog box:

    -Enter a significance level, such as 0.05 or 0.01.
    -Enter the number of runs.
    -Enter the number of elements of the first type.
    -Enter the number of elements of the second type.

5. Click on the **Evaluate** bar.

6. Click on **Plot** to display a graph with the test statistic and critical values.

The textbook includes an example involving Boston rainfall amounts for Monday; those values are listed in Data Set 17 of Appendix B in the textbook. Letting D represent a dry day (with 0 rainfall) and letting R represent a rainy day (with a positive amount of rainfall), we get the sequence shown below. We want to determine whether rain on Mondays is random.

DDDDRDRDDRDDRDDDRDDRRRDDDDRDRDRRRDRDDDRDDDDRDRDDRDDDR

The textbook describes the procedure for examining the above sequence to find these results:

n_1 = number of Ds = 33

n_2 = number of Rs = 19

G = number of runs = 30

Having found the number of elements of each type and the number of runs, we can use STATDISK to obtain the results shown in the display on the next page.

Runs Test for Randomness

Significance, α: 0.05

Num Runs: 30

Num Element 1:   33

Num Element 2:   19

Using Approximation
Mean, μ_G 25.115
St Dev, σ_G 3.3061
Test Statistic, z   1.4775
Critical z   ±1.9600

Fail to Reject the Null Hypothesis

Data does not provide enough
evidence to indicate the sequence
is not random

Evaluate   Help   Plot

The above STATDISK display includes the value of $\mu_G = 25.115$, $\sigma_G = 3.3061$, the test statistic of $z = 1.4775$, and the critical values of $z = -1.96$ and $z = 1.96$. These calculations are somewhat messy, as shown below.

$$\mu_G = \frac{2n_1 n_2}{n_1 + n_2} + 1 = \frac{2(33)(19)}{33 + 19} + 1 = 25.115$$

$$\sigma_G = \sqrt{\frac{(2n_1 n_2)(2n_1 n_2 - n_1 - n_2)}{(n_1 + n_2)^2 (n_1 + n_2 - 1)}}$$

$$= \sqrt{\frac{(2)(33)(19)[2(33)(19) - 33 - 19]}{(33 + 19)^2 (33 + 19 - 1)}} = 3.306$$

$$z = \frac{G - \mu_G}{\sigma_G} = \frac{30 - 25.115}{3.306} = 1.4775$$

STATDISK easily provides the results for these difficult calculations. Based on the results, we fail to reject the null hypothesis of randomness. It appears that on Mondays in Boston, the occurrence of rain is a random event.

CHAPTER 13 EXPERIMENTS: Nonparametric Statistics

13-1. **Qwerty vs. Dvorak Keyboards** Exercise 2 in Section 13-2 of *Elementary Statistics* uses word ratings for the Qwerty and Dvorak keyboard configurations. Using the sentence of "The quick brown fox jumped over the lazy red dog," we get Qwerty word ratings of 2, 5, 7, 3, 5, 5, 2, 3, 2, 1. The corresponding Dvorak ratings are 0, 5, 5, 3, 5, 3, 0, 4, 1, 1. We want to determine whether there is sufficient evidence to support the claim that the Dvorak keyboard has lower ratings. Use STATDISK's *sign test* with a 0.05 significance level, and test the claim that there is no significant difference between the two keyboard configurations. Enter the results below.

Test statistic:_____   Critical value:_____

Conclusion: _____

_____

13-2. **Effect of Transformation** When dealing with inconvenient sample data, it is often helpful to transform them by doing such things as subtracting a constant from each value or multiplying each value by a constant. **Repeat** Experiment 13-1 after multiplying each value by 100 and enter the results below.

Test statistic:_____   Critical value:_____

Conclusion: _____

_____

How were the sign test results affected by the multiplication of each value by 100?

_____

_____

How would the sign test results be affected if we add the same constant to each sample value?

_____

_____

How would the sign test results be affected if a single value is changed to become an outlier?

_____

_____

13-3. **Parametric vs. Nonparametric**   Do Experiment 13-1 by using a parametric test as described in Chapter 8 of this manual/workbook. Compare the parametric t test results to the sign test results obtained in Experiment 13-1. Do the results lead to the same conclusion? Is either test more sensitive to the differences between the pairs of data?

_____

_____

13-4. **Sign Test vs. Wilcoxon Signed-Ranks Test**   Repeat Experiment 13-1 by using the Wilcoxon signed-ranks test for matched pairs. Enter the STATDISK results below, and compare them to the sign test results obtained in Experiment 13-1. Specifically, how do the results reflect the fact that the Wilcoxon signed-ranks test uses more information?

Test statistic:_____   Critical value:_____

Conclusion: _____

_____

Comparison: _____

_____

13-5. **Effects of Transformations and Outliers**   Design and conduct experiments to answer the following.

Given a sample of matched data, if a constant is added to (or subtracted from) each value, how are the Wilcoxon signed-ranks test results affected?

_____

Given a sample of matched data, if every sample value is multiplied (or divided) by a nonzero constant, how are the Wilcoxon signed-ranks test results affected?

_____

Given a sample of matched data, if one sample value is incorrectly entered as an exceptionally large value, how are the Wilcoxon signed-ranks test results affected?

_____

13-6. **Sign Test for Median**   Exercise 6 from Section 13-2 of the textbook refers to the weights of sugar in Domino packets. Those sample values are listed in Data Set 4 in Appendix B and they are stored in the STATDISK file named sugar.sdd. Use STATDISK to determine whether the median amount of sugar in the packets is equal to 3.5 oz.

Test statistic:_____   Critical value:_____

Conclusion: _____

_____

13-7. **Wilcoxon Rank-Sum vs. Parametric Test** Section 8-2 of this manual/workbook includes a STATDISK display for a parametric test of the claim that weights of regular Coke and regular Pepsi are different. The sample statistics are reproduced here. Use STATDISK with the Wilcoxon rank-sum test and compare the results to those found with the parametric test. Specifically, how do the nonparametric test results reflect the fact that the ranks of the data are used instead of the actual sample values? (The sample values are in STATDISK files ckregwt.sdd and ppregwt.sdd, and they are listed in Data Set 1 of Appendix B in the textbook.)

	Regular Coke Weight (pounds)	Regular Pepsi Weight (pounds)
n	36	36
$\bar{x}$	0.81682	0.82410
s	0.007507	0.005701

Rank sum for regular Coke:_____   Rank sum for regular Pepsi:_____

Test statistic: _____   Critical values: _____

Conclusion:_____

_____

Comparison of Tests: _____

_____

13-8. **Effects of Transformations and Outliers** Refer to the sample data and results given in Experiment 13-7. Design and conduct experiments to answer the following.

In a collection of two independent samples, if a constant is added to (or subtracted from) each sample value, how are the Wilcoxon rank-sum results affected?

_____

In a collection of two independent samples, if every sample value is multiplied (or divided) by the same nonzero constant, how are the Wilcoxon rank-sum test results affected?

_____

In a collection of two independent samples, if one sample value is incorrectly entered as an exceptionally large value and becomes an outlier, how are the Wilcoxon rank-sum test results affected?

_____

13-9. **Kruskal-Wallis Test vs. ANOVA**   Refer to the crash dummy car test results listed in Table 11-1 in Section 11-1 of this manual/workbook. Use STATDISK's Kruskal-Wallis test to test the claim that the *chest deceleration* measurements are the same for the four different car size classes. Then compare those results to the results that are obtained using analysis of variance. Specifically, how do the results reflect the fact that the Kruskal-Wallis test is based on the ranks of the sample data whereas analysis of variance uses the actual sample values?

Test statistic: _____   Critical values: _____

Conclusion:_____

_____

Comparison of Tests: _____

_____

13-10. **Does It Rain More on Weekends?**   Data Set 17 in Appendix B of *Elementary Statistics* lists the amounts of rainfall (in inches) for each day of the week in a recent year. Those amounts are stored in the STATDISK files rainmon.sdd, raintues.sdd, rainwed.sdd, rainthur.sdd, rainfri.sdd, rainsat.sdd, and rainsun.sdd. Use The Kruskal-Wallis test to determine whether the rainfall amounts appear to be different on different days of the week.

Test statistic: _____   Critical values: _____

Conclusion:_____

_____

13-11. **Effect of an Outlier** Repeat Experiment 13-10 after changing one of the sample values to 500 in. How are the Kruskal-Wallis results affected by the presence of an outlier?

Test statistic: _____   Critical values: _____

Conclusion:_____

_____

Effect of an outlier:_____

_____

13-12. **Illustrating Test Sensitivity**   Construct three sets of sample data with five values in each sample. Assume that we want to test the null hypothesis (at the 0.05 significance level) that the samples come from the same population. Arrange the data so that the Kruskal-Wallis test conclusion is failure to reject the null hypothesis, while analysis of variance (Section 11-2 of the textbook) leads to rejection of the null hypothesis.

Sample 1: _____
Sample 2: _____
Sample 3: _____

13-13. **Relationship Between Salary and Physical Demand** Exercise 4 from Section 13-6 of *Elementary Statistics* includes paired salary and stress level *ranks* for 10 randomly selected jobs. The physical demands of the jobs were also ranked; the salary and physical demand ranks are given below (based on data from *The Jobs Rated Almanac*). Does there appear to be a relationship between the salary of a job and its physical demands?

Salary	2	6	3	5	7	10	9	8	4	1
Physical demand	5	2	3	8	10	9	1	7	6	4

Test statistic: _____   Critical values: _____

Conclusion: _____

_____

Why should we *not* use the linear correlation coefficient with the given sample values?

_____

13-14. **Cigarette Nicotine, Tar, Carbon Monoxide** Data Set 8 in Appendix B of *Elementary Statistics* lists sample measurements of cigarette tar, nicotine, and carbon monoxide. Those data sets are stored in STATDISK with the file names of tar.sdd, nicotine.sdd, and co.sdd. Use STATDISK's rank correlation for the following.

Use the paired data consisting of tar and nicotine and test for a correlation.

Test statistic: _____   Critical values: _____

Conclusion: _____

_____

Use the paired data consisting of carbon monoxide and nicotine and test for a correlation.

Test statistic: _____   Critical values: _____

Conclusion: _____

_____

Researchers would like to develop a method for predicting the amount of nicotine, and they want to measure only one other item. In choosing between tar and carbon monoxide, which is the better choice? Why?

_____

13-15. **Small Samples and Large Samples** Exercises 8 and 16 from Section 13-7 of *Elementary Statistics* both list sequences of World Series wins by American League and National League teams. In both exercises, recent results are given with American and National league teams represented by A and N, respectively. In both exercises, we want to test the null hypothesis that the given sequence is random. If you solve both exercises using manual calculations, how do the methods of solution differ? How do the methods of solution differ when you use a software package such as STATDISK?

Small sample data (from Exercise 8 in Section 13-7 of the textbook):
A N A N A A A N N A A N N N N A A A N A N A N A A A N A N A

Test statistic: _____   Critical values: _____

Conclusion: _____

_____

Large sample data (from Exercise 16 in Section 13-7 of the textbook):
A N A N N N A A A A N A A A A N A N N A A N N A A A A N
A N N A A A A A N A N A N A N A N A A A A A A A A N N A N A N
N A A N N N A N A N A N A A A A N N A A N N N N A A A A N A
N A N A A A N A N A

Test statistic: _____   Critical values: _____

Conclusion: _____

_____

Comparison of methods with manual calculations and with STATDISK:

_____

_____

13-16. **Stock Market: Testing for Randomness Above and Below the Median** Exercise 10 in Section 13-7 of the textbook notes that trends in business and economics applications are often analyzed with the runs test. The accompanying list shows (in order by row) the annual high points of the Dow Jones Industrial Average for a recent sequence of years. Using STATDISK, first find the median of the values, then replace each value by A if it is above the median and B if it is below the median. Then use STATDISK to apply the runs test to the resulting sequence of As and Bs.

969	842	951	1036	1052	892	882	1015	1000	908
898	1000	1024	1071	1287	1287	1553	1956	2722	2184
2791	3000	3169	3413	3794	3978	5216	6561	8259	9338

Test statistic: _____   Critical values: _____

Conclusion: _____

_____

What does the result suggest about the stock market as an investment consideration?

_____

13-17. **Mark McGwire Homeruns** Exercise 15 in Section 13-7 of the textbook refers to the homerun distances for baseball player Mark McGwire. Those distances are listed in Data Set 19 from Appendix B of the textbook, and they are also listed in the STATDISK file mcgwire.sdd. Use STATDISK to test for randomness above and below the median distance.

Test statistic: _____   Critical values: _____

Conclusion: _____

_____

13-18. **Random Number Generator** STATDISK can randomly generate sample values from different populations. Select **Data**, then **Uniform Generator**, then proceed to generate 100 random numbers from a uniformly distributed population of whole numbers (0 decimal places) with a minimum of 1 and a maximum of 500.

First *explore* the generated data set by using STATDISK to investigate center, variation, distribution, and outliers. List the relevant observations and the statistical tools used to form them:

_____

_____

_____

Use STATDISK to determine whether the generated numbers appear to be random. Identify the criterion used as the basis for your test.

Test statistic: _____   Critical values: _____

Conclusion: _____

_____

Criterion: _____

_____

Index

STATDISK's Menu Configuration

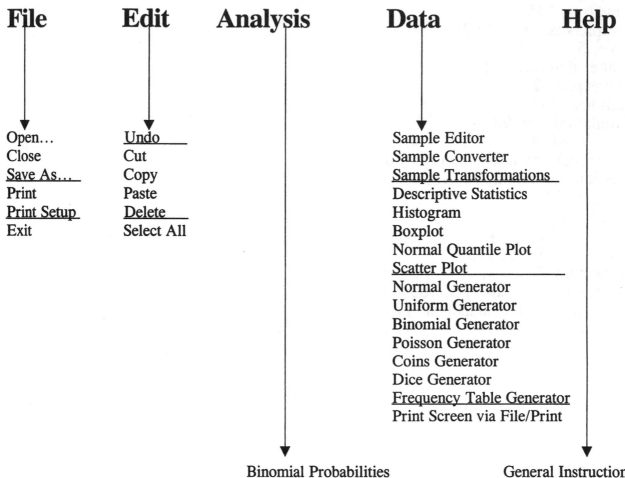

File

Open…
Close
Save As…
Print
Print Setup
Exit

Edit

Undo
Cut
Copy
Paste
Delete
Select All

Analysis

Data

Sample Editor
Sample Converter
Sample Transformations
Descriptive Statistics
Histogram
Boxplot
Normal Quantile Plot
Scatter Plot
Normal Generator
Uniform Generator
Binomial Generator
Poisson Generator
Coins Generator
Dice Generator
Frequency Table Generator
Print Screen via File/Print

Help

Binomial Probabilities
Sample Size Determination
Confidence Intervals
Hypothesis Testing
Correlation and Regression
Multiple Regression
Multinomial Experiments
Contingency Tables
One-Way Analysis of Variance
Sign Tests
Wilcoxon Tests
Kruskal-Wallis Test
Rank Correlation
Runs Test
Probability Distributions

General Instructions
About STATDISK